Ball Lightning as a Messenger from a New World of Circulating Light

Ball Lightning as a Messenger from a New World of Circulating Light

By

V. P. Torchigin and A. V. Torchigin

Cambridge
Scholars
Publishing

Ball Lightning as a Messenger from a New World of Circulating Light

By V. P. Torchigin and A. V. Torchigin

This book first published 2023

Cambridge Scholars Publishing

Lady Stephenson Library, Newcastle upon Tyne, NE6 2PA, UK

British Library Cataloguing in Publication Data
A catalogue record for this book is available from the British Library

ISBN (10): 1-5275-5068-0
ISBN (13): 978-1-5275-5068-1

TABLE OF CONTENTS

ACKNOWLEDGEMENTS

The book was prepared as a part of the State assignment of the Russian Academy of Sciences No. 0069-2019-0010.

PREFACE

This book describes a theory of the finally unraveled nature of ball lightning and the consequences of this event for physics in the field of the interaction of light with matter. The description is based on more than eight dozen articles published since 2003 in leading international journals in physics and optics. The Introduction presents convincing arguments that ball lightning is completely different from what was assumed by the numerous researchers who have put forward more than 500 different hypotheses and theories about the nature of ball lightning so far. It is shown that ball lightning is an object whose existence in nature no one suspected, and after its description, few can agree that such an object can exist.

This object is so simple that its essence can be described in one short sentence. This is a bubble, the spherical shell of which is a thin film of highly compressed air, in which ordinary white light circulates in all possible directions. Since 2003, in numerous publications, we have shown that the behavior of such a bubble in the Earth's atmosphere, in accordance with the generally accepted laws of physics and optics, completely coincides with the mysterious, intriguing and inexplicable behavior of natural ball lightning.

We consider a unique object that can detect a flying airliner, catch up and accompany it, even though the object should be blown away by airflow, the speed of which is several times higher than in the strongest hurricane. It can enter rooms through window panes. It can move upwind. It can find holes and cracks and penetrate them while changing its shape if necessary. If a science fiction writer described such an object, then the reader's reaction would be you can fantasize, but you need to know when to stop.

Taking into account that the behavior of the bubble is exactly the same as that of ball lightning, and applying the well-known duck test to the bubble - "If it looks like a duck, swims like a duck and quacks like a duck, then it is probably a duck" - we can say that this bubble is probably a ball lightning.

It remains to explain the inexplicable - how can such a light bubble exist in nature? This is the subject of this book.

At the same time, it turned out that the existing knowledge about the interaction of light with matter is incomplete. This concerns the magnitude of the momentum of light in an optical medium, the reasons for its increase compared to the momentum of light in free space, the magnitude of the forces acting on the optical medium in which light propagates, and the magnitude of the pressure created by light in various media, including gases. Therefore, before proceeding to the justification of the stability of the bubble, it is necessary to have a clear understanding of the laws on the basis of which such a justification is carried out.

Initially, the laws that are used to determine the necessary conditions for the existence of the bubble are presented. As for the experimental confirmation of the existence of such bubbles, nature has created experimental samples, and numerous eyewitness accounts can be considered as documentation of the observed properties of such bubbles. So the experiments have already been carried out and their results coincide with theoretical predictions.

After the laws of physics and optics necessary for stability analysis are presented, which are consistent with other generally accepted laws, it is shown that the stability of the bubble arises with an extremely high compression of air in the bubble shell and an extremely high light energy density. Other parameters of the bubble were also determined.

No modern knowledge about quantum mechanics, the theory of relativity, nuclear physics, thermodynamics, etc. is required. Knowledge at the level of the 18-19th century is sufficient for understanding. The book can be used as a textbook on optics and physics in the field of the interaction of light with matter.

This is the only book in the world in which for the first time the correct physical theory of a ball lightning is presented and the physical phenomena responsible for its appearance and anomalous behavior are considered. The owners of the book will witness that they lived at a time when the centuries-old mystery of ball lightning was solved and a new stage in the development of mankind began.

INTRODUCTION

No great discovery was ever made without a bold guess.
I. Newton

Piotr Kapitsa, a Nobel Prize winner, studied the phenomenon of ball lightning, and at the end of his life was forced to admit that this nut was too strong for his teeth. At the same time, he assumed that ball lightning was a phenomenon from another world unknown to him. Indeed, he was right. There is a whole world in which circulating light plays a major role. Ball lightning is one of the most famous and mysterious representatives of this world, which we call the world of circulating light.

This statement implies that the nature of the light in ball lightning is well understood. It really is. Since 2003, when the first publication about the completely unexpected nature of ball lightning appeared (Torchigin 2003, 108-116), more than eight dozen articles have been published in leading international journals in physics and optics, which not only explain the physical nature of ball lightning, but also show that the mysterious, intriguing and paradoxical behavior of ball lightning in the earth's atmosphere has a very simple explanation.

Typically, the Introduction describes the previous state of the problem discussed in the book. The phenomenon called ball lightning has been known for several centuries. Currently, more than 500 different theories and hypotheses about the nature of this phenomenon are known, and several thousand articles have been published. But the mysterious and intriguing behavior of ball lightning, which is described in many books and databases, remains unexplained. As the author of one of the scientific books about ball lightning said out of desperation: "All theories are united by one thing, they do not work" (Sagan 2004). The same author complained that: how can one describe the physical nature of ball lightning if it violates the basic laws of physics?

Since we are spared the need to describe the existing theories about the nature of ball lightning, we can describe the physical nature of ball lightning immediately in this Introduction, since the structure of ball lightning is very simple, and its description takes only a few lines. Ball lightning is a completely unusual physical object. Previously, no one suspected the existence of such objects. Therefore, it could not be studied either theoretically or experimentally. Of course, this object does not have a name.

The essence of this object is easiest to explain when comparing it with the well-known soap bubble. Unlike a soap bubble, the shell of this object consists of a thin spherical layer of strongly compressed air where conventional white light is circulating in all possible directions. The light compresses the air to minimize the total energy of the light and compressed air. The reflective index of the compressed air becomes greater than that of the surrounding air. In this case, the shell is a spherical lightguide that prevents radiation of light in free space. Thus, our object is a symbiosis of two elements. These are the air and light. Compressed air ensures the existence of the circulating light, and the circulating light ensures the existence of compressed air. Our object has nothing to do with electricity. Therefore, in order to distinguish references to our imaginary object from references to natural ball lightning, we cut off the tail in the word "lightning" and will call our object ball light or bubble light or circulating light.

This short explanation can be clarified a bit. The air pressures inside and outside the spherical layer of compressed air are the same and equal to atmospheric pressure. Unlike a soap bubble, whose spherical shape is provided by the excess air pressure inside the soap film, the spherical shape of our object is provided by the centrifugal force produced on the layer by the light circulating in it. The glow of ball lightning is explained by the fact that the circulating light is gradually scattered due to the molecular light scattering. Similarly, the sunlight is scattered in the earth's atmosphere. The blue color of the sky and orange color of the Sun near horizon are explained by the light scattering in the earth's atmosphere.

In these two paragraphs, we have fully described in detail the physical nature of ball lightning. In the next paragraphs, we explain the reasons for its abnormal behaviour. Since the 17th century, it has been known that the light propagating in an inhomogeneous optical medium is deflected in the direction where the reflective index increases. Thus, the circulating light also moves in the direction where the reflective index increases. Using more understandable physical parameters, we can say that the circulating light moves in the direction of increased terrestrial air density. If we take into account that the air pressure is the same in the region where the ball lightning is usually observed, then the air density can be various if the air temperatures in this region are various. The lower the air temperature, the higher the air density.

Thus, the behaviour of ball light obeys the simplest rule. Ball light tends to move in the direction of decreasing air temperature. However, in some cases, when moving over long distances, the ball light moves in the direction of increasing air density. This situation occurs when a ball lightning falls from the sky to the earth or catches up with a flying airliner that disturbs the uniform distribution of air density in such a way that the maximum air density is at the leading edge of the fuselage or wings of the airliner.

Observational properties of ball lightning are given in books by authors such as Singer (1971), Stakhanov (1979), Barry (1980), and Steinhoff (1998), where more than a thousand of the most revealing events are described in detail. Physicists cannot imagine an object obeying accepted laws of physics, the properties of which would coincide with the observed properties of ball lightning.

There is a legitimate question. What are the grounds for asserting that the presented nature of ball lightning in the form of circulating light corresponds to reality, while numerous other theories are erroneous? Nature made sure that the answer to this question was convincing. Nature provided ball lightning, which has a set of unique and anomalous properties that no other object known to date has had. The exception is our object, which has the same paradoxical properties as natural ball lightning. Here are some of the most mysterious.

First, the anomalous property of ball lightning to penetrate into rooms through window panes ought to be taken into account. This wonderful property is confirmed by 46 similar cases described in Grigor'ev, Grigor'eva, and Shiryaeva (1992, 279). It is known that no particles, not even electrons, can penetrate glass. What then does ball lightning consist of? It is difficult to imagine a theory that states that ball lightning consists of nothing.

Second, it is the anomalous property of ball lightning to move in a direction that does not coincide with the direction of the wind. A section entitled "Flies against wind" in Sagan's 2004 book, based on the declassified ball lightning information from the Oak Ridge National laboratories in the USA, is devoted to a description of this property. Sagan rightly notes that it is difficult to create a theory out of an object that violates the basic laws of physics.

Third, ball lightning has the ability to catch up and accompany a flying airliner. An entire chapter entitled "Airplane Crashes" is devoted to a description of this phenomenon in Sagan (2004). In this case, ball lightning is not blown away by the wind, whose speed exceeds the wind speed in the strongest hurricane.

Bearing in mind these features, we can conclude that none of the known hypotheses satisfies these requirements, since it is difficult to imagine a hypothesis in which it is stated that ball lightning consists of nothing and violates the basic laws of physics.

Such unusual, mysterious properties of ball lightning have a great advantage. They allow one to easily reject various theories and hypotheses about the nature of ball lightning because the objects proposed as ball lightning do not have the same abnormal properties that natural ball lightning has. On the other hand, if our ball light has the same set of anomalous properties as natural ball lightning, we have good reason to assert that ball light and ball lightning are one and the same.

The question may arise - what kind of object can penetrate a window pane without damaging it, because air cannot penetrate through the glass? The answer is simple. Air does not penetrate glass, only light does. On the other side of the glass is just such air. The penetrating light compresses this air. For light, glass is just another optical medium with a different reflective index.

In our first paper, published in 2003, we showed that the behavior of such circulating light in the terrestrial atmosphere completely coincided with the behavior of natural ball lightnings, based on numerous eyewitness accounts. Our explanation was based on the simplest well-known laws of physics and optics. Later, we explained in more detail all the anomalies of natural ball lightnings in numerous publications.

Another argument in favor of the fact that ball light is ball lightning can be the so-called duck test, by which it is possible to identify an unknown object by observing its properties and characteristic behavior. The test sounds like this: "If it looks like a duck, swims like a duck, and quacks like a duck, then it's probably a duck."

We will apply this test to the Ball Light (the unknown subject which properties were considered in the articles) and compare its properties and behavior with that of the natural Ball Lightning (duck), the behavior and characteristics of which are well known from numerous evidences of eyewitnesses. The following similarities ought to be noted.

Both Ball Lightning and Ball Light have a spherical form.

Both Ball Lightning and Ball Light radiate white light, but their temperature is essentially smaller than that of a usual body-radiated white light.

Both Ball Lightning and Ball Light have lifetimes ranging from a few seconds to a few minutes (Torchigin 2019, 2961).

Both Ball Lightning and Ball Light disappear unexpectedly (Torchigin 2019, 65-69).

Both Ball Lightning and Ball Light disappear instantly (Torchigin 2019, 65-69).

Both Ball Lightning and Ball Light disappear without traces (Torchigin 2019, 65-69).

Both Ball lightning and Ball Light bypass obstacles (Torchigin 2018, 167).

Both Ball Lightning and Ball Light fall from sky but do not strike the earth's surface (Torchigin 2018, 149).

Both Ball Lightning and Ball Light can find out holes and slots to penetrate within buildings (Torchigin 2019, 163126.).

Both Ball Lightning and Ball Light change their form to penetrate through small holes and narrow slots (Torchigin and Torchigin 2016, 6155).

Both Ball Lightning and Ball Light can penetrate into rooms through window panes (Torchigin and Torchigin 2016, 5876).

Both Ball Lightning and Ball Light can move against wind (Torchigin 2019, 533-537).

Both Ball Lightning and Ball Light can catch up to a flying airliner and penetrate inside its saloon (Torchigin and Torchigin 2017, 196).

Both Ball Lightning and Ball Light can burn out metal objects (Torchigin 2021, 167528).

In accordance with the duck test, since both Ball Lightning and Ball Light have identical behavior and characteristics, Ball Light is probably Ball Lightning. Thus, our ball light passes the duck test.

Furthermore, it has been shown that circulating light exists not only in ball lightning. There is a whole world of luminous objects with anomalous properties in which circulating light is the cause of their anomaly. We can say that there is a hitherto unknown world of circulating light, inhabited by objects with anomalous properties. The most famous of this world of circulating light is ball lightning. Unlike ball lightning, which is a rare natural phenomenon, other objects from the world of circulating light are obtained in laboratories, usually when trying to get artificial ball lightning. The properties of these objects are well studied, which allows a deeper understanding of the nature of ball lightning. The presence of circulating light in such objects makes it possible to explain their anomalous properties.

Having made sure that the properties and behavior of a bubble of light and ball lightning in the earth's atmosphere completely coincide, we faced a more difficult problem - to show, on the basis of existing knowledge, that the bubble of light could be stable. This was a task where the positive answer known in advance, but for two decades no theoretical physicist could indicate the physical laws and phenomena responsible for the existence of bubbles of light and their stability.

Essentially, compressed air and circulating light both had a tendency to expand, and it was not clear why their interaction should prevent these trends.

We eventually managed to find a solution to this problem (Torchigin 2019, 704; Torchigin 2020, 165098; Torchigin 2021, 166635). The study of the necessary conditions under which stability can exist has led to results according to which the energy densities of compressed air and circulating light should be record highs. Based on the stability conditions, other parameters of the circulating light were determined (Torchigin 2021, 167390). It turned out that the lifetime of circulating light was approximately three orders of magnitude longer than the lifetime of sunlight in the Earth's atmosphere. This conclusion is also supported by observations.

We present the reasons for the arising of stability of the circulating light and the conditions for its occurrence. Skeptics may argue that the paper tolerates everything, but where is the evidence for the correctness of the presented theory? Any theory must be experimentally confirmed. Indeed, we present only a model of ball lightning. Any model only approximately describes the real phenomenon and requires experimental confirmation. Fortunately, nature presents numerous experimental samples in the form of natural ball lightnings and a complete set of information about their properties in the form of numerous eyewitness accounts. From this information, it follows that experimental samples exist. Moreover, these samples have exactly the same set of anomalous properties (inexplicable on the basis of modern knowledge). It is shown that the properties of ball lightning are not anomalous, but are in full accordance with the generally accepted and long-known basic laws of optics and physics.

The great scientist Faraday, after whom the unit of electrical capacity is named, said: "Nothing is too wonderful to be true if it is consistent with the laws of nature." The famous American writer Mark Twain also respected reality. He believed that "Fiction is obliged to stick to possibilities. Truth isn't. "

To substantiate the stability of circulating light, it is necessary to have reliable knowledge about the interaction of light with matter. Unfortunately, there is no such generally accepted knowledge. This applies to almost all issues necessary to determine the stability, such as the magnitude of the momentum of light in the optical medium, the magnitude of the forces acting on the light from the optical medium in which it propagates, and the pressure exerted by light on the air in which it circulates. Without reliable answers to these questions, the study of the stability of circulating light loses its meaning. Therefore, before considering the stability of circulating light, a theory of the interaction of light with matter is presented, which is of independent value and can be used in other applications. It should be immediately noted that the indisputable generally accepted laws of physics and optics without any additional assumptions in all the considerations given below were used. In other words, it can be argued that the results below are a direct consequence of generally accepted laws. In this regard, the book can be seen as a textbook that outlines the current understanding of the above issues.

First of all, this refers to the change in the magnitude of the momentum of light in the optical medium relative to the momentum of the same light in free space. There are irrefutable arguments that the momentum of light in an optical medium increases and equally irrefutable arguments that the momentum of light in an optical medium decreases. Without a reasonable idea of the magnitude of the momentum of light in compressed air, further reasoning loses its meaning.

Since the forces arising at an interaction between light and matter following Newton's laws are determined by the derivative of the momentum over time, any conclusion regarding these forces can be doubted under the pretext that there is no generally accepted idea of the magnitude of the momentum. The first chapter deals with the question of the magnitude of the momentum of light in an optical medium. Regarding the magnitude of the momentum in free space, there is a generally accepted opinion. However, up to now there has been no generally accepted opinion on the magnitude of momentum of light in an optical medium. There is not even a qualitative understanding. The question of increasing or decreasing the momentum of light in an optical medium in comparison with the momentum of light in free space is still being debated.

Without going into this discussion, we have shown that the momentum of light in an optical medium is increased by a factor of n, where n is the reflection index of the optical medium. This conclusion is confirmed by a dozen thought experiments published in leading international journals. Such a result existed before the dispute arose and followed directly from the laws of mechanics. Particularly convincing arguments have been obtained in the analysis of the momentum of other types of waves, such as elastic waves and sound waves. In this case, the momentum of an electromagnetic wave is a special case of a more general law. This makes the results more reliable.

However, another thought experiment, by Balazs, also based on the generally accepted laws of physics and optics, remained inexplicable, from which it follows that the momentum of light in an optical medium decreases compared to the momentum of light in free space. The second chapter is devoted to the reconciliation of these contradictions. The modified thought experiment of Balazs is considered. In contrast to the original experiment, where a light wave was considered, a wave of an arbitrary type was considered. It turned out that the conclusions drawn from the results of this experiment should be interpreted differently. It is shown that the above reconciliation can be achieved under the assumption that additional forces must act on the homogeneous medium in which the wave propagates. For the particular case of an electromagnetic wave, these forces coincide in magnitude with the Abraham forces, the existence of which has not yet ceased disputes. An experimental confirmation of the existence of these forces for an ultrasonic wave propagating in the air is given.

The nature of the momentum of an arbitrary type wave is considered. An explanation is given as to why the momentum of a body with mass is proportional to its velocity, while the momentum of a wave is inversely proportional to its velocity. It is shown that the wave entails the material medium in which it propagates. It is the movement of the material medium in which the wave propagates that ensures the existence of a momentum for the wave. On the assumption that an electromagnetic wave propagating in free space is not an exceptional case, then it should be recognized that such a wave entrains free space (or tries to entrain, creating pressure on it).

The third chapter is devoted to the analysis of the physical nature of the forces produced by waves of various types propagating in a homogeneous medium. Since the existence of these forces is still being debated, it can be assumed that these unknown forces are new forces that can complement knowledge about the nature of waves. It is shown that these forces have a different nature depending on the type of energy carried by a particular wave. Formulas are given for calculating these forces for waves of various physical natures, such as an elastic wave, a sound wave, a torsion wave, a wave in a stretched string, and a sea wave in shallow water. Particular attention was paid to the study of such forces produced by an electromagnetic wave. Formulas are given for such forces produced by an electromagnetic wave in a dielectric. These formulas are compared with similar formulas obtained on the basis of a perfectly different approach using the Lorentz forces, which is currently popular due to the lack of a generally accepted opinion on the magnitude of the momentum of light in matter. It is shown that the formulas have significant differences. A special case is considered to show that the formulas based on Lorentz forces give an incorrect result. Corrections are given to which these formulas must be subjected.

The fourth chapter is devoted to the investigation of the phenomenon that provides compression of the air in the shell of circulating light. This phenomenon was discovered by Helmholtz in 1888 for the dielectric placed in an electrical field. The same phenomenon takes place for an optical medium where the light wave is propagating. In this case, the phenomenon is called optical electrostriction pressure and is considered as one of the possible mechanisms of optical nonlinearity. It is shown that, contrary to the generally accepted concept, this phenomenon manifests itself completely differently in gas than in liquid and solids. Furthermore, it was shown that an increase in the air reflective index could occur not only with an increase in the air density but also with the absorption into the shells of other gases, the reflective indexes of which exceed the air reflective index.

The fifth chapter is devoted to the justification of the stability of circulating light. First of all, the necessary conditions are derived under which the circulating light in a spherical shell of compressed air is stable. Since the stability of such objects has not been previously studied, special attention was paid to substantiating the research methodology. For this purpose, the chosen methodology was initially tested on objects whose stability is not in doubt. These were a soap bubble and a children's balloon. The necessary conditions for the stability of these objects were determined.

After that, exactly the same methodology was used to determine the necessary conditions for the stability of the circulating light. At the same time, immediately before the start of calculations, it turned out that stability can occur at a sufficiently high value of the reflective index of compressed air, which can only be achieved with its extremely high compression. It is shown that the steady state and its stability appear only at the very great air pressure produced by the very great energy intensity of the circulating light.

From these necessary conditions, the parameters of the circulating light were derived, such as stored energy, compressed air pressure, the energy density of circulating light, the lifetime of circulating light, and features of its occurrence and disappearance. The last chapter is devoted to an analysis of the processes connected with the arising, existence and disappearance of the circulating light.

It can be concluded that the presented optical theory of ball lightning makes it possible not only to solve the riddle about the nature of ball lightning, the solution of which no-one suspected, but also to gain new knowledge about the interaction of light with matter.

Almost 20 years have passed since convincing evidence was presented that the behaviours of a bubble of light and a natural ball lightning were the same, and therefore a bubble of light existed in nature. It would seem that the discovery of a new natural phenomenon should have aroused interest in it and in the study of the presence of circulating light in other natural phenomena. However, this did not happen.

Having an advantage over other researchers who only doubted and shook their heads about the existence of circulating light in nature, we, based on the belief that circulating light exists, not only showed its stability, but also began to deduce new knowledge from this fact. We analyzed the interaction of waves of arbitrary types with the mediums in which they propagate, and obtained new results that are valid not only in the context of circulating light, but also in a general case. The book expands our knowledge of nature.

The information presented in the book could be accessible to scientists of the 19th century, when concepts about the reflective index of an optical medium *n*, momentum, force, mass, and energy appeared. True, it is necessary to know the foundations of the mathematical analysis developed by Leibniz and Newton at the same time. We tried to present the book in such a way that the presentation of the new theory of ball lightning was combined with the presentation of modern ideas about the foundations of modern optics, in part concerned the relationship of optical and mechanical phenomena. Therefore, the book can be considered a textbook on the fundamentals of modern optics, which were partially corrected and supplemented during the development of the theory of ball lightning.

This book is the first book where the theoretical grounds that explain an existence of natural ball lightnings and the laws responsible for their existence are presented.

CHAPTER ONE

MOMENTUMS OF ARBITRARY WAVES

Nothing is too wonderful to be true, if it be
consistent with the laws of nature.
M. Faraday

The purpose of this section is to prove that the momentum of light in an optical medium is increased by a factor of n as compared with the momentum of the same light in free space where n is the reflective index of the medium.

The momentum of light in free space is well known, but there is still no generally accepted idea of the magnitude of the momentum of light in an optical medium. There is still no generally accepted opinion about whether the momentum of light in an optical medium increases or decreases compared to the momentum of the same light in free space. With the Minkowski approach, the momentum in the optical medium increases by a factor of n. In the Abraham approach, the momentum in the optical medium decreases by the same factor.

Currently, there is an opinion that both approaches are applicable in different situations. For example, we can read Griffith (2012, 7): "Beginning in the late 1960's something approaching a consensus emerged: Both the Minkowski momentum and the Abraham momentum are "correct," but they speak to different issues, and it is largely a matter of taste which of the two one identifies as the "true" electromagnetic momentum. Correct both forms".

Another compromise was proposed by Mansuripur (2010, 1997-2005), according to whom the momentum of a light wave in an optical medium was the arithmetic average of the values proposed by Minkowski and Abraham, that is, it was equal to $(n + 1/n)/2$.

Most recently, Barnett and Loudon reanalyzed the controversy and argued that both momenta are "correct" because both can be measured, but in different situations (Barnett and Loudon 2010, 927). There is also the Balazs thought experiment (Balazs 1953, 411), the interpretation of which shows that the Abraham approach is correct. Following Balazs' analysis and repeating arguments of his thought experiment, they concluded that "it is difficult to see how any component of our derivation could seriously be open to question … If argument advanced in favour of the Abraham momentum were to be incorrect, then that would bring into question uniform motion of an isolated body as expressed in Newton's first law of motion".

Lengthy attempts to derive the value of optically induced force (OIF) from the Lorentz law for forces acting on a moving charge in electric and magnetic fields are known. These attempts began with the work of Einstein and Laub (1908), resumed in the work of Gordon (1973, 14), and continued more recently in the works of Laudon and Mansuripur (Mansuripur, Zakharian, and Wright 2013, 023826; Loudon and Barnett 2006, 11855). Various formulas for a magnitude of optically induced forces (OIF) have been obtained, but not one of them gives the correct result in the simplest particular case. We do not give here arguments related to the criticism of various incorrect expressions for calculating OIF based on Lorentz forces. The correct expression, in our opinion, is presented in Brevik (2009, 219301).

Such a situation with the momentum of light in the optical medium cannot be considered satisfactory. We cannot develop a theory of circulating light using concepts for which there is no generally accepted recognition. In this case, the theory may cause reasonable doubts about its correctness. Below, we present evidence that the momentum of light in an optical medium increases by n times.

First, we calculate the momentum of light in the optical medium for cases where the mention of electricity is not required. Moreover, the magnitude of the momentum is derived for an abstract wave, where a mention of the reflective index is absent, as in the case of the Doppler effect. Then we will show, as an explanation and illustration, that the same result can be obtained if we take into account the expressions known since Faraday's era for forces acting on a dielectric in a constant electric field.

We present arguments based on various physical phenomena that the momentum of a light pulse in the optical medium corresponds to the Minkowski approach and is greater by n times than that in free space, where n is the reflective index of the optical medium. We then refute a single opposite argument in the form of the Balazs thought experiment, according to which the Abraham approach is correct and the momentum of light in the optical medium decreases by n times. Thus, after a dozen arguments presented in favour of the Minkowski approach, we conclude that the age-old debate about the magnitude of the momentum of light in an optical medium should be completed in favour of the Minkowski approach.

Thus, we consider the debate about the magnitude of the momentum of light in the optical medium complete (at least in the context of where the forces act on the light due to a change in its momentum). By the way, a correct magnitude of the momentum of light in the optical medium could be derived from results of Thomson who determined in 1904 the pressure produced by the light at a normal incidence on a flat boundary between the optical medium and free space (Thomson 1904). The knowledge of the magnitude of the momentum of light in an optical medium allows one to calculate the optically induced forces applied to the circulating lights and to calculate their motion without the knowledge that light is an electromagnetic wave.

The momentum of a wave of arbitrary physical nature

The concept of momentum refers not only to bodies and electromagnetic waves, but also to other waves for which such physical concepts as the speed of propagation, frequency, and energy density exist. We believe that the calculation of the momentum in the general case for waves of an arbitrary physical nature allows us to calculate the momentum for electromagnetic waves. In this case, long disputes about the momentum of the electromagnetic wave remain set aside, as disputes about particular cases, when the result in a general case is known.

Notation and basic definitions

First of all, it is worth noting that our notion of the momentum of light was taken from mechanics, where Newton introduced this concept long before a discussion arose about the magnitude of the momentum of light in an optical medium. For a body of mass m moving at speed v, the momentum of the body is defined as follows

$$\boldsymbol{p} = m\,\boldsymbol{v} \tag{1.1}$$

Following Newton, the momentum characterizes the "quantity of motion". A reason for a change of the momentum p is force F that in accordance with the second Newton law is given by

$$\boldsymbol{F} = \frac{d\boldsymbol{p}}{dt} \tag{1.2}$$

In a closed system (one that does not exchange any matter with the outside and is not acted on by outside forces) the total momentum is constant. This fact, known as the law of conservation of momentum, is implied by Newton's laws of motion.

A magnitude of the momentum of a light pulse propagating in a vacuum is generally accepted and is given by

$$M_0 = \mathscr{E} / c \tag{1.3}$$

where $\mathscr{E}$ is the energy of the light pulse.

If a continuous plane light wave is considered, any notion of the energy is senseless, and the energy density W of the wave should be considered. Analogously, the notion of the momentum for a continuous wave is senseless also and the momentum flux density (MFD) that is equal to the momentum of the wave propagating through a cross-section of the unit area per unit time is considered. In other words, MFD is equal to the momentum of the fragment of the continuous wave, the cross-section of which is equal to the unit area, and the length of which is equal to the distance that the wave travels per unit time. The MFD of a continuous plane light wave propagating in a vacuum at speed c is given by $(W_0c)/c=W_0$ [J/m^3], where W_0 is the energy density of the light wave in free space.

For continuous light wave, the density force F in Eq. (1.2) is substituted by the pressure P, and dp/dt is substituted by a difference between input and output MFDs. As a result, the pressure P produced by the wave propagating in an optical medium is determined as follows

$$P = MFD_{IN} - MFD_{OUT} \tag{1.4}$$

For example, a mechanical pressure P applied to a reflector by the light wave at a normal reflection from the reflector in accordance with Eq. (1.4) is equal

$$P = MFD_{IN} - MFD_{OUT} = MFD_{IN} - (-MFD_{IN}) = 2MFD_{IN} \tag{1.5}$$

Thus the pressure produced on the reflector at the normal reflection of a continuous wave, for which MFD is equal to MFD_{IN}, is equal to twice MFD_{IN}.

When we talk about waves in an optical medium and free space, we mean the MFD of this wave. Our immediate goal is to convince the reader of the correctness of the Minkowski approach with the help of numerous thought arguments.

Optically induced forces (OIF) are produced by the light propagating in an inhomogeneous optical medium. As a result, the light interacts with matter (an exchange of the momentums between the light and matter takes place). The law of the conservation of momentum and the third Newton law are valid at this interaction. As a result, each OIF changes the mechanical MFD of matter. In turn, a counterpart of the OIF that arises in accordance with the third Newton law changes the MFD of light. Thus, each interaction is accompanied by redistribution between the mechanical momentum of the optical medium and the momentum of light. The sum of these momentums is not changed.

We ought to note that the dimensions of the following three physical values: energy density W [J/m^3], pressure P [N/m^2=Nm/m^3=J/m^3], and momentum density flux MFD [$\frac{J}{m/s}\frac{m/s}{m^3}=\frac{J}{m^3}$] are identical. The difference is that MFD is a vector and W is a scalar. On the assumption that the MFD in an optical medium that increases by n times, the magnitudes of the MFD and W in an optical medium are identical also because W increases in an optical medium by n times.

It should be noted that the above concepts (force, energy density, momentum flux density, pressure, and wave propagation velocity) are also applicable to waves of a different physical nature (in particular, elastic waves). There is no dispute about the magnitude of the momentum for other types of waves. For waves of any physical nature for which the above concepts are applicable, the MFD of this wave can be determined. After that, the momentum for an electromagnetic wave can be considered as a special case, and the debate will end on this. We begin our consideration with examples valid for waves of various physical natures.

The momentum of arbitrary waves derived from the Doppler law

In accordance with the Doppler law (Eden 1992), an increase of the frequency of the wave reflected from the reflector that moves towards the wave is given by

$$\Delta\omega = \omega\frac{2v_R}{v_W} \tag{1.6}$$

where ω is the frequency of the incident wave, and v_W and v_R are the velocities of the incident wave and reflector, respectively. Let us consider a fragment of the wave of unit area and length v_W and energy $\mathcal{E}$. The momentum of this fragment is equal to the MFD of the wave. The volume V of the fragment is equal to v_W.

For definiteness, we restrict ourselves to considering light and sound waves. Taking into account that the energy of a photon or phonon is proportional to its frequency, we can conclude from Eq. (1.6) that

$$\Delta\mathcal{E} = \mathcal{E}\frac{2v_R}{v_W} = 2Wv_R \tag{1.7}$$

where

$$W = \mathcal{E} / v_W \tag{1.8}$$

is the energy density of the wave. This fragment is reflected from the reflector at a time interval 1s. At this time the reflector moves at distance v_R. The mechanical effort that is spent at this displacement is given by

$$\Delta\mathcal{E} = Pv_R \tag{1.9}$$

where P is the pressure produced by the wave at the reflection. Comparing Eq. (1.7) and Eq. (1.9), we have

$$P = 2W \tag{1.10}$$

The *MFD* of the fragment changes its sign at the reflection. The pressure due to the change of the momentum of the wave is given by

$$P = 2MFD \tag{1.11}$$

Comparing Eq. (1.10) and Eq. (1.11) we can see that

$$MFD = W \tag{1.12}$$

Thus, the MFD of an arbitrary wave is equal to the energy density of the wave. As is seen from Eq. (1.8), the energy density is in inverse proportion with the wave speed.

Concepts and terms inherent for waves of an arbitrary physical nature were used. The light waves can be considered as a special case. We believe that the overall result is more convincing than the specific one.

In the particular case of the light wave, the energy density of the fragment in an optical medium is greater by n times than that in free space. In this case, the MFD in an optical medium in accordance with Eq. (1.12) and Eq. (1.8) is greater by n times than that in free space.

The momentum of an arbitrary wave derived from its adiabatic invariant

Let us compare two plane resonators consisting of two ideal reflectors, as is shown in Fig.1-1. The first resonator is located in the first medium, and the distance between reflectors is equal *to* $\lambda/2$, where λ is the wavelength in the medium. The second resonator is located in the second medium, where the distance between reflectors is equal to $\lambda/(2\gamma)$ and where $\gamma>1$ is the ratio of speeds of waves in the resonators. In this case, the eigen frequencies of the resonators are identical.

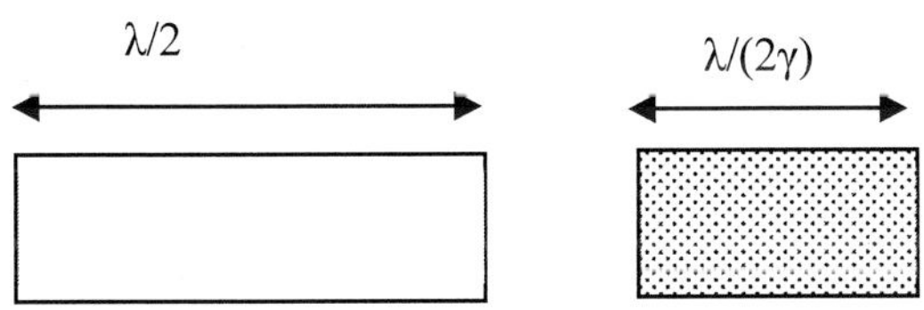

Fig.1-1. Comparison of two plane resonators with identical eigen frequencies.

Let the energy of oscillations also be identical. Let us decrease the lengths of the resonators by one percent. The eigen frequency increases by one percent also. Following the theory of oscillations of lossless conservative isolated oscillations systems, there is the adiabatic invariant under which a change of the energy stored in the resonator is proportional to a change in its eigen frequency at a slow change of any resonator parameters. We can then conclude that the energies stored in the resonators increase by one percent in both resonators.

The mechanical works that are then required to decrease the distance between the reflectors are identical for both resonators. But a change in the distance between reflectors for the second resonator is γ times smaller than for the first one. To have identical mechanical works, the force applied to the reflector should be greater by γ times for the second resonator than for the first one. Since the force overcomes the pressure produced by the wave at the reflection from the reflector, the pressure in the second case is greater by γ times than in the first one. Since pressures are produced by a change of the MFD of the wave at the reflection, the MFD of the wave in the second resonator is greater by γ times than that in the first one. Thus, the MFDs of waves differ by γ times, provided that their energies are identical. We can also conclude that the MFD and energy density of wave in the second resonator is greater by γ times than in the first one.

Since the presented arguments are true for waves of an arbitrary physical nature, they are also true for electromagnetic waves that can be considered as a special case. Assuming that the first medium is free space, and the second medium is an optical medium with a reflective index $n = \gamma$, we can conclude that the MFD of light in an optical medium increases by $\gamma=n$ times. We believe that the overall result is more convincing than the specific one.

The momentum of an arbitrary wave derived from the balance of its momentums at refraction

Let us consider the refraction of any wave at a boundary between two mediums where the same wave propagates at speeds v_1 and v_2. The incidence angle is α_1 and the refracted angle is α_2, as shown in Fig. 1-2..Let any part of the wave penetrate from medium 1 to medium 2. The horizontal component of the momentum is preserved on the condition given by

$$p_1 sin\alpha_1 = p_2 \sin \alpha_2 \tag{1.13}$$

where p_1 and p_2 are MFDs of waves in mediums 1 and 2, respectively.

The periodicities of these waves at the boundary should be identical. This condition can be written as follows

$$k_1 \sin \alpha_1 = k_2 \sin \alpha_2 \tag{1.14}$$

where k_1 and k_2 are propagation constants in the first and second mediums and

$$k_1 = \omega / v_1, \; k_2 = \omega / v_2 \tag{1.15}$$

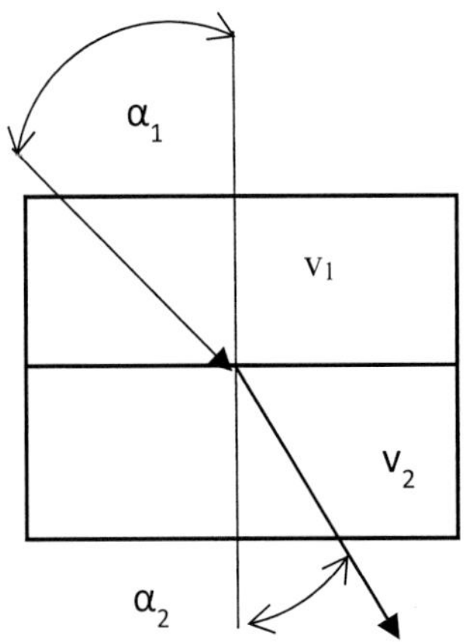

Fig.1-2. Propagation of a wave of an arbitrary physical nature through a boundary of two mediums.

From Eq. (1.14) and Eq. (1.15), we have

$$\frac{\sin\alpha_1}{v_1} = \frac{\sin\alpha_2}{v_2} \tag{1.16}$$

Taking into account Eq. (1.13) and Eq. (1.16), we have

$$\frac{p_1}{p_2} = \frac{v_2}{v_1} \tag{1.17}$$

Setting the light wave at $v_1 = c$, $v_2 = c/n$, we see that the MFD of light in an optical medium increases by n times as compared with that in free space from simple geometric consideration.

The momentum of a wave derived from the analysis of a circulating wave

Let a wave of a whispering gallery type circulate around the equator of the ball. The circulating wave contains an integer number of wavelengths, i.e. the phase shift along the equator is given by

$$2\pi Rk - 2\pi N \tag{1.18}$$

where the propagation constant $k=\omega/v$, ω is the angular frequency, v is the velocity of the wave within the ball, R is the radius of the ball, and N is any integer. Eq. (1.18) preserves at a gradual change the parameters R and k. In this case, we have from Eq. (1.18)

$$\omega R / v = \omega_0 R_0 / v_0 \tag{1.19}$$

where ω_0, v_0, R_0 are parameters in any particular case. Furthermore, it is known that the energy $\mathcal{E}$ stored in an isolated conservative oscillating system is proportional to the eigen frequency ω at a slow change of the parameters of the system. In this case, we have $\mathcal{E} R = const$ or

$\ln \mathcal{E} + \ln R = const$ or $\frac{d\mathcal{E}}{\mathcal{E}} + \frac{dR}{R} = 0$ or

$$\frac{d\mathcal{E}}{dR} = -\frac{\mathcal{E}}{R} \tag{1.20}$$

The term $\frac{d\mathcal{E}}{dR}$ in Eq. (1.20) designates the total force F (a sum of modules of all surface forces) applied to the surface of the ball. The total force tends to increase the radius of the ball. Eq. (1.20) can then be rewritten as follows

$$F = \frac{\mathcal{E}}{R} \tag{1.21}$$

This force is equal to the centrifugal force applied to the entire surface of the ball. The force is produced by circulating light, the total momentum of which is equal to p. Following Newton's second law, the force F is given by

$$F = \frac{dp}{dt} = p\Omega \tag{1.22}$$

where Ω is the angular frequency of the circulating wave around the equator and is given by

$$\Omega = 2\pi \frac{v}{2\pi R} = \frac{v}{R} \tag{1.23}$$

From Eq. (1.22) and Eq. (1.23), we have

$$F = p\frac{v}{R} \tag{1.24}$$

Comparing Eq. (1.21) and Eq. (1.24), we have

$$p = \frac{\mathcal{E}}{v} \tag{1.25}$$

From Eq. (1.25) we can conclude that the momentum of light in free space p_0 is given by

$$p_0 = \frac{\mathcal{E}}{c} \tag{1.26}$$

Since the velocity of the light in an optical medium decreases by n times, from Eq. (1.25) and Eq. (1.26) we see that the momentum in an optical medium p increases by n times as compared with the momentum p_0 in free space that corresponds to the Minkowski approach.

The ratio of the momentums of an arbitrary wave propagating in various media

Let us consider the refraction of any wave at a boundary of two mediums, as shown in

Fig.1-3, where velocities of the wave are different and are equal to v_1 and v_2, respectively. Let the incident wave be given by

$$a = a_0 \exp[j(\omega t - \boldsymbol{k} \cdot \boldsymbol{r})] \tag{1.27}$$

Here, the propagation constant is characterized by the vector $\boldsymbol{k}$ that is directed along the direction of propagation of the incident wave. The module of the vector is given by

$$k = \omega / v_1 \tag{1.28}$$

where $\omega=2\pi f$, and f is the frequency of the wave.

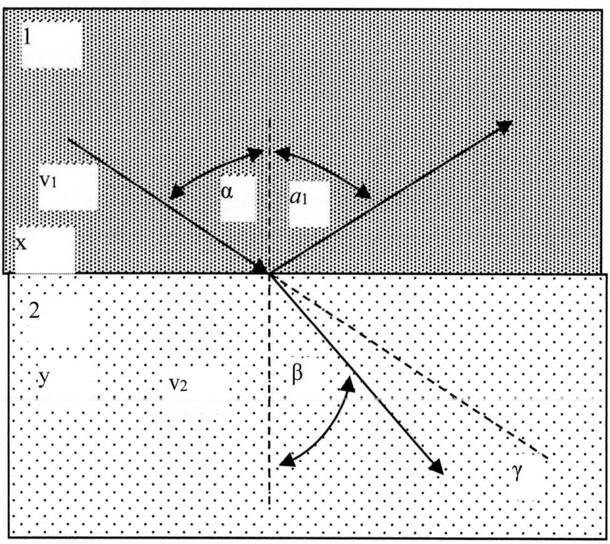

Fig.1-3. Penetration of a part of an incident wave from medium 1 to medium 2.

Let the x-axis be parallel to the projection of vector $\boldsymbol{k}$ on the boundary plane. If the wave reflects completely from the boundary, the x-component of the momentum of the wave is preserved because the angle of incidence is equal to the angle of reflection that is $\alpha=\alpha_1$, and $p_1\text{Cos}\alpha=p_1\text{Cos}\alpha_1$ where p_1 is the momentum of the incident wave.

The situation is changed if a part of wave $\gamma<1$ penetrates into medium 2 where the speed of the wave changes from v_1 to v_2. The x-component of the momentum of the wave is preserved if the following condition is valid

$$p_1\gamma \sin\alpha = p_2\gamma \sin\beta \tag{1.29}$$

where β is the angle of refraction.
The reflective wave cannot propagate at the same angle α, as is shown in

Fig.1-3 by the dotted line, because the momentum p_2 of the wave in medium 2 is different from that in medium 1.

We do not know the relation between the momentums in mediums 1 and 2, but we can determine relations between angles α and β from a simple geometric consideration.

The space period Λ_1 along the x-axis that is connected with the incident wave can be presented as follows

$$a = a_0 \exp[j(\omega t - \frac{2\pi x}{\Lambda_1})] \tag{1.30}$$

At the same time, taking into account Eq. (1.27) and Eq. (1.28), the same space period is given by

$$a = a_0 \exp\{j[\omega t - kxCos(\pi/2 - \alpha)]\} = a_0 \exp[j(\omega t - \frac{\omega}{v_1} xSin\alpha)] \tag{1.31}$$

Comparing space periods in Eq. (1.30) and Eq. (1.31), we have

$$\Lambda_1 = \frac{2\pi v_1}{\omega Sin\alpha} \tag{1.32}$$

Repeating the same considerations for the wave in medium 2, we obtain that the space period along the x-axis that is connected with wave 2 is given by

$$\Lambda_2 = \frac{2\pi v_2}{\omega Sin\beta} \tag{1.33}$$

Space periods Λ_1 and Λ_2 should be identical. Then, from Eq. (1.32) and Eq. (1.33) we have

$$\frac{v_1}{v_2}=\frac{Sin\alpha}{Sin\beta} \tag{1.34}$$

From Eq. (1.29) and Eq. (1.34), we can conclude that the *x*-component of the momentum is unchanged if

$$\frac{p_1}{p_2}=\frac{v_2}{v_1} \tag{1.35}$$

In the particular case of the light wave, $v_1=c$ is the velocity of light in free space, $v_2=c/n$ is the velocity of light in an optical medium, and in accordance with Eq. (1.35) we have

$$\frac{p_1}{p_2}=\frac{1}{n} \tag{1.36}$$

This means that Minkowski was right and the momentum of light in an optical medium increases by *n* times.

The momentum of an electromagnetic wave derived from the laws of electrostatics

What force affects light? In general, can force act on the light? Where is the region of the light beam at which the force acts? Where is the object that is responsible for this force located? We believe most readers have never heard of this force. We will then pose the question differently. Can light create a force that will act on the optical medium in which it propagates? The answer to this question is yes. Indeed, since the time of Maxwell, it has been known that a dielectric placed in an electric field is under the action of a force. Everyone can check this phenomenon. If you bring an electrified comb closer to a small piece of paper, this piece will be attracted to the comb.

It does not matter if the comb is positively or negatively charged. In any case, a piece of paper will be attracted to the comb. In this case, it can be argued that a piece of paper will be attracted to the comb if the charge on it will periodically change from positive to negative and vice versa. In other words, a piece of paper is subject to the force produced by an alternate electric field. The direction of this force is perpendicular to the surface of the paper.

As was established at the end of the 19th century, the light is an electromagnetic wave in which there is an alternate electric field. Thus, a force must act on the dielectric perpendicular to its surface. In optics, transparent dielectrics are used, called optical media. Instead of opaque pieces of paper, inhomogeneities of the optical medium are used.

Thus, a dielectric located in a static electrical field is subjected to the action of the density force, and the force density applied to the dielectric is given by

$$f = -\varepsilon_0 grad(\varepsilon)\frac{E^2}{2} \tag{1.37}$$

where ε is the permittivity of the dielectric and E is the strength of the electric field (*Physical encyclopedia*, 1992). On the other hand, the energy density W_E of the electrical field is given by

$$W_E = \varepsilon_0 \varepsilon E^2 / 2 \tag{1.38}$$

As is known, the energy densities of both electrical and magnetic fields are identical in an electromagnetic wave. In this case, the total energy density W_{EH} is two times greater than the energy density of the electrical field and is given by

$$W_{EH} = \varepsilon_0 \varepsilon E^2 \tag{1.39}$$

From Eq. (1.37) and Eq. (1.39) we have

$$f = -grad(\varepsilon)\frac{W_{EH}}{2\varepsilon} \tag{1.40}$$

Taking into account the following relation, $\varepsilon = n^2$, where n is the reflective index of the optical medium, we have

$$f = -grad(n)\frac{W_{EH}}{n} \tag{1.41}$$

Taking into account that the energy density of an electromagnetic wave depends on n, the following condition is valid

$$\frac{W_{EH}}{n} = \frac{W_{EH0}}{n_0} \tag{1.42}$$

Here W_{EH0} and n_0 are the energy density and the reflective index in free space, respectively. Then, from Eq. (1.41) we have

$$f = -grad(n)W_{EH0} = -grad(n)W_0 \tag{1.43}$$

Here the energy density of an electromagnetic wave in free space is denoted by W_0.

Let a light wave propagate in an inhomogeneous optical medium along the x axis, as shown in Fig.1-4. The reflective index n increases slowly along the x axis so that $dn/dx = grad(n) << 1/\lambda$ where λ is the wavelength of the wave. In this case, reflections of the wave can be neglected.

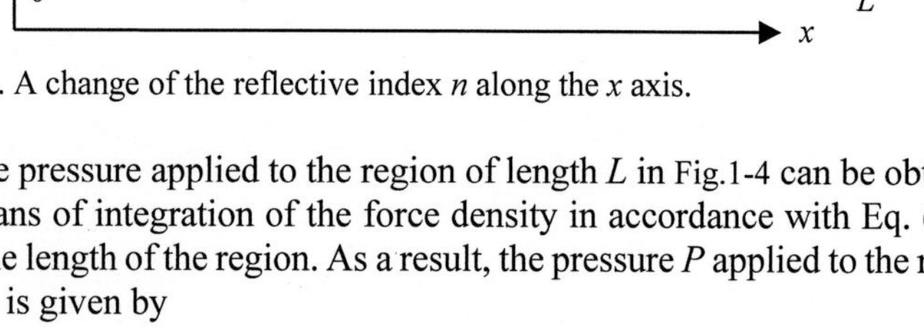

Fig.1-4. A change of the reflective index n along the x axis.

The pressure applied to the region of length L in Fig.1-4 can be obtained by means of integration of the force density in accordance with Eq. (1.43) over the length of the region. As a result, the pressure P applied to the region $0<x<L$ is given by

$$P = \int_0^L -grad(n)W_0 = -W_0[n(L) - n(0)] = -W_0(n-1) \tag{1.44}$$

The initial MDF in free space is equal to the wave energy density W_0. Taking into account that $MDF_{IN}=W_0$, we have from Eq. (1.44) that the MDF in the optical medium can be calculated from the following condition

$$-(n-1)W_0 = W_0 - MFD \tag{1.45}$$

From Eq. (1.45) we have

$$MFD = nW_0 \tag{1.46}$$

Thus, MDF in an optical medium increases by n times as compared with that in free space, and in accordance with Eq. (1.42) is equal numerically to the energy density W of the light wave in the optical medium.

Thus, we have shown that the MFD in an optical medium increases by n times by means of analysis of MFDs in the region, the pressures on which are calculated by independent and different approaches based on the density forces in an inhomogeneous dielectric.

The momentum of light derived from the geometric optics

As has been known since Euler's time, a light beam that is propagating in an inhomogeneous optical medium is deflected from a straight line, and its trajectory is determined by laws of the geometrical optics in accordance with the following eikonal equation

$$\frac{d}{ds}(n\boldsymbol{e}_s) = grad(n) \tag{1.47}$$

where $\boldsymbol{e}_s$ and s are the unit vector along the tangent of the trajectory and the length along the trajectory, respectively.

Let the momentum of a light pulse propagating in an optical medium be equal to

$$\boldsymbol{p}_m = \frac{\mathcal{E} n\boldsymbol{e}_s}{c} . \tag{1.48}$$

Let us determine the force applied to the light pulse from the optical medium if the pulse is propagating along with $grad(n)$. Then, from Eq. (1.47) and Eq. (1.48) we have

$$\frac{d\boldsymbol{p}_m}{dt} = \frac{d\boldsymbol{p}_m}{ds}\frac{ds}{dt} = grad(n)\frac{\mathcal{E}}{c}\frac{c}{n} = \frac{grad(n)}{n}\mathcal{E} \tag{1.49}$$

On the other hand, in accordance with the law of electrostatics, the force applied to the optical medium from the side of the light pulse is given by Eq. 1.41. We can see that the right-hand terms in Eq. (1.42) and Eq. (1.49) differ by signs only. This is in accordance with Newton’s third law, which states that the force applied to the light pulse is opposite to the force applied to the optical medium.

Thus, knowing that the eikonal equation Eq. (1.47) and the force applied to the optical medium by the light pulse are in accordance with the law of electrostatics given by Eq. (1.42), we can obtain Eq. (1.49) and therefore Eq. (1.48).

We can say that Newton's second law of motion for a light pulse is derived from the eikonal equation Eq. (1.47) and the law of electrostatics Eq. (1.42). The eikonal equation can be derived from the Snell law through simple mathematical manipulation. In turn, the Snell law can be derived from simple geometric considerations where any mentions of forces and momentums are absent. Thus we have derived the second Newton’s second law for the propagation of light in an inhomogeneous optical medium from the law of electrostatics. Using the laws of Newton and electrostatics and performing transformations in the opposite direction, we get Eq. (1.48).

Since the velocity of the pulse is equal to c/n, Eq. (1.48) can be presented as follows

$$\boldsymbol{p} = \frac{\mathcal{E}\boldsymbol{e}_s}{v} \tag{1.50}$$

Thus, the trajectory of a light pulse in an inhomogeneous optical medium is determined by the same Newtonian law that he discovered three centuries ago for various bodies. Newton’s second law is valid for light pulses provided that the momentum of the pulse is proportional to the reflective index n in the place where it is propagating.

The momentum of light derived from the Maxwell stress tensor

Let a light pulse of unit cross-section and energy density W_0 and length L propagate in free space. Its momentum is known and is equal to

$$p_0 = \frac{\mathcal{E}}{c} = \frac{W_0 L}{c} \tag{1.51}$$

The energy of the light pulse in an optical medium is preserved, but the energy density becomes nW_0, and its length becomes L/n. Then the force applied to the pulse in accordance with Eq. (1.42) is given by

$$F = grad(n)W_0\frac{L}{n} = \frac{dn}{dx}W_0\frac{L}{n} = \frac{dn}{dt}\frac{dt}{dx}W_0\frac{L}{n} = \frac{d(W_0 n)}{dt}\frac{n}{c}\frac{L}{n} = \frac{d(LW_0 n/c)}{dt} \tag{1.52}$$

Following Newton's second law, we have

$$F = \frac{dp}{dt} \tag{1.53}$$

From Eq. (1.51), Eq. (1.52), and Eq. (1.53) we have

$$p = \frac{nW_0 L}{c} = np_0 \tag{1.54}$$

We can see that the momentum p in an optical medium increases by n times as compared with the momentum of the light pulse p_0 of the same wave in free space.

This conclusion can be derived at the time when it becomes known that the nature of light is an electromagnetic wave and a momentum of light pulse in free space is given by $\mathcal{E}/c$.

Propagation of light in an inhomogeneous optical medium

Let the inhomogeneity of the medium be such that the permittivity ε and permeability μ change only along the z-axis. The density force applied to the medium is given by

$$f = -\varepsilon_0 grad(\varepsilon)E^2/4 - \mu_0 grad(\mu)H^2/4 \tag{1.55}$$

where E and H are amplitudes of the electric and magnetic fields in the medium, respectively (*Physical encyclopedia*, 1992).

If a light pulse of length L and cross-section S is propagating in the medium, it produces in the medium density forces under Eq. (1.55). In accordance with Newton's third law, the force applied to the pulse is given by

$$F = LS(\varepsilon_0 grad(\varepsilon)E^2 / 4 + \mu_0 grad(\mu)H^2 / 4) \quad (1.56)$$

Taking into account that the energy density of the electrical field is given by

$$W_E = \varepsilon_0 \varepsilon E^2 / 4 \quad (1.57)$$

and the energy density of the magnetic field is given by

$$W_H = \mu_0 \mu H^2 / 4 \quad (1.58)$$

Eq. (1.56) can be presented as follows

$$F = LS[\varepsilon_0 / \varepsilon)grad(\varepsilon)W_E + (\mu_0 / \mu)grad(\mu)W_H] \quad (1.59)$$

Taking into account that W_E=W_H in an electromagnetic wave and ε=1 and μ=1 in free space, Eq. (1.59) can be presented as follows

$$F = LS[\frac{grad(\varepsilon)}{\varepsilon} + \frac{grad(\mu)}{\mu}]W_{EH} / 2 = LS\frac{grad(\varepsilon\mu)}{\varepsilon\mu}W_{EH} / 2 = LS\frac{grad(\sqrt{\varepsilon\mu})}{\sqrt{\varepsilon\mu}}W_{EH} \quad (1.60)$$

where W_{EH} is the total energy density.

Let a light pulse of total energy $\mathcal{E} = LCW_{EH}$ propagate along the direction of $grad(\sqrt{\varepsilon\mu})$, and this direction coincides with the z-axis. Then, Eq. (1.60) is transformed as follows

$$F = \mathcal{E}\frac{grad(\sqrt{\varepsilon\mu})}{\sqrt{\varepsilon\mu}} = \frac{\mathcal{E}}{\sqrt{\varepsilon\mu}}\frac{d(\sqrt{\varepsilon\mu})}{dt}\frac{dt}{dx} = \frac{\mathcal{E}}{\sqrt{\varepsilon\mu}}\frac{d(\sqrt{\varepsilon\mu})}{dt}\frac{\sqrt{\varepsilon\mu}}{c} = \frac{\mathcal{E}}{c}\frac{d(\sqrt{\varepsilon\mu})}{dt} \quad (1.61)$$

At the same time, the momentum of the light pulse in free space is given by

$$p_0 = \frac{\mathscr{E}}{c} \tag{1.62}$$

Taking into account Eq. (1.61) and Eq. (1.62), we can conclude that

$$F = \frac{dp_M}{dt} \tag{1.63}$$

where

$$p_M = p_0\sqrt{\varepsilon\mu} \tag{1.64}$$

Concluding remarks

Assuming that the electromagnetic wave is no exception among other types of waves, we first determined, on the basis of thought experiments, the momentum of an arbitrary wave, about which there is no dispute. It is shown that the momentum flux density of a wave propagating in an inhomogeneous medium is inverse to the propagation velocity of this wave. In this case, the momentum flux density is equal to the energy density of the wave. As applied to an electromagnetic wave, it follows that momentum flux density increases by a factor of n compared with that in free space.

Several thought experiments that take into account the specifics of the electromagnetic wave are then presented. In particular, it is shown that from the laws of electrostatics and magneto statics, with respect to the forces acting on a matter located in electric and magnetic fields, it directly follows that the momentum flux density of an electromagnetic wave during its transition from free space to the matter increases by n times, where n is a generalized reflective index, equal to the ratio of the wave velocity in free space to that in the matter.

From all the above thought experiments based on the generally accepted laws of physics and optics, without prior assumptions, it follows that the momentum of light in the optical medium increases by n times. However, according to Balazs's thought experiment, which is also based on the generally accepted laws of physics and optics, the momentum of light in an optical medium decreases by a factor of n. This contradiction is resolved in the next chapter.

CHAPTER TWO

FORCES PRODUCED BY WAVES IN HOMOGENEOUS MEDIUMS

Truth is ever to be found in the simplicity, and not in the multiplicity and confusion of things.
I. Newton

The Balazs thought experiment is based on the generally accepted laws of physics and optics (Balazs 1953, 195341), from which it follows that the momentum of light waves in an optical medium decrease as compared with those in free space. This result contradicts our conclusions obtained from the above thought experiments, and without reconciliation of all thought experiments, it is impossible to determine which of the thought experiments correspond to reality.

The essence of this experiment is as follows. An isolated closed system is considered, consisting of a medium in which a light pulse propagates at speed c and a transparent block of the optical medium through which this pulse propagates at a lower speed v. Two cases are considered. In the first case, it is assumed that the light pulse passes outside the block at a speed c. In the second case, it is assumed that the pulse passes through the block at speed $v<c$. Since the speed of the pulse inside the block is less, the pulse will travel a shorter distance at the same time, that is, the pulse will be closer to the block than in the first case. Furthermore, it is assumed that the pulse has a mass. Then, in the second case, the center of the mass will be closer to the block than in the first case, that is, the center of the mass of the system will shift against the direction of wave propagation.

Since, in accordance with the laws of mechanics, the center of mass of an isolated system cannot be changed under the action of internal forces, it is concluded that the block must move in the direction of wave propagation. This means that the light pulse drags the block along with it when it is propagating inside the block, that is, the pulse transfers part of its momentum to the block. It follows from the law of conservation of momentum that inside the block the momentum of the light pulse decreases. The importance of this experiment is connected with the fact that neither of the preliminary notions about optically induced forces (OIF) - their physical nature and the phenomena responsible for their arising - are used.

Most recently, Barnett and Loudon reanalyzed the arguments of the Balazs thought experiment and concluded that "it is difficult to see how any component of our derivation could seriously be open to question … If argument advanced in favor of the Abraham momentum were to be incorrect, then that would bring into question uniform motion of an isolated body as expressed in Newton's first law of motion".

To overcome these contradictions, a modification of the Balazs thought experiment was developed, in which the concept of the mass of an optical pulse was not used at all (Torchigin and Torchigin 2016, 6120-6128). It turns out that the conclusions deduced due to the Balazs thought experiment can be interpreted in a perfectly different way, which allows us to discover new types of forces, the existence of which were unknown until now.

Modified Balazs thought experiment

The scheme of the modified thought experiment is shown in Fig. 2-1. Four ideal reflectors are mounted on a cart that can move without friction in a horizontal direction in such a way that a pulse of wave propagating in a medium at velocity v_1 is reflected from each reflector at an angle of 90 degrees. The reflectors and cart are an indivisible whole. The pulse is radiated towards reflector 1. When the pulse is reflected from this reflector, the pulse acquires a positive momentum in the horizontal direction. In accordance with the law of conservation of momentum, the cart acquires a negative mechanical momentum of the same magnitude, that is, it begins to move in a negative direction with a certain velocity, as shown in the upper right graph.

The cart moves until the pulse is reflected from reflector 2. In this case, the horizontal component of the pulse momentum disappears and, accordingly, the horizontal component of the momentum of the cart disappears also, that is, the cart stops and stands until the pulse is reflected from reflector 3. At the same time, a negative horizontal component of the momentum arises, and the cart acquires a positive momentum, that is, the cart begins to move in a positive direction, as shown in the same graph.

The cart moves until the pulse is reflected from reflector 4. In this case, the horizontal component of the pulse momentum disappears and, accordingly, the horizontal component of the momentum of the cart disappears, that is, the cart stops. Since the times of movement of the cart in the positive and negative directions are the same, after the disappearance of the pulse, the system returns to its original position and the center of mass of the system remains unchanged.

A different situation takes place if block 6 of length L is inserted on the path of pulse propagation from reflector 3 to reflector 4. The velocity of the pulse propagation decreases from v_1 to v_2 within the block. Let the duration of the pulse be smaller than the time of the pulse propagation within the block as is shown in Fig. 2-1. In this case, the pulse propagation time increases in accordance with the expression

$$\Delta t = \left(\frac{L}{v_2} - \frac{L}{v_1}\right) = L\left(\frac{1}{v_2} - \frac{1}{v_1}\right) \tag{2.1}$$

Assuming that the velocity of the cart does not change while the pulse propagates inside the block, we obtain that the cart will not stop in its original position, but will travel further, by the distance determined by the expression

$$\Delta s = \frac{p}{m}\Delta t = \frac{pL}{m}\left(\frac{1}{v_2} - \frac{1}{v_1}\right) \tag{2.2}$$

Here p/m is the velocity of the cart where p is a value of the cart momentum, and m is the total mass of the cart, reflectors, wheels, etc. In order for the cart to be in its original position, it is necessary that during the time $T{=}L/v_2$ while the pulse is propagating inside the block, the cart moves the distance Δs, i.e., so that its negative speed is determined by the expression

$$v = \frac{\Delta s}{T} = \frac{\Delta s}{L/v_2} = \frac{p}{m}\frac{1/v_2 - 1/v_1}{1/v_2} = \frac{p}{m}\left(1 - \frac{v_2}{v_1}\right) \tag{2.3}$$

The cart then acquires additional positive momentum p_{Cart}, as given by

$$p_{\mathrm{Cart}} = mv = p\left(1 - \frac{v_2}{v_1}\right) \tag{2.4}$$

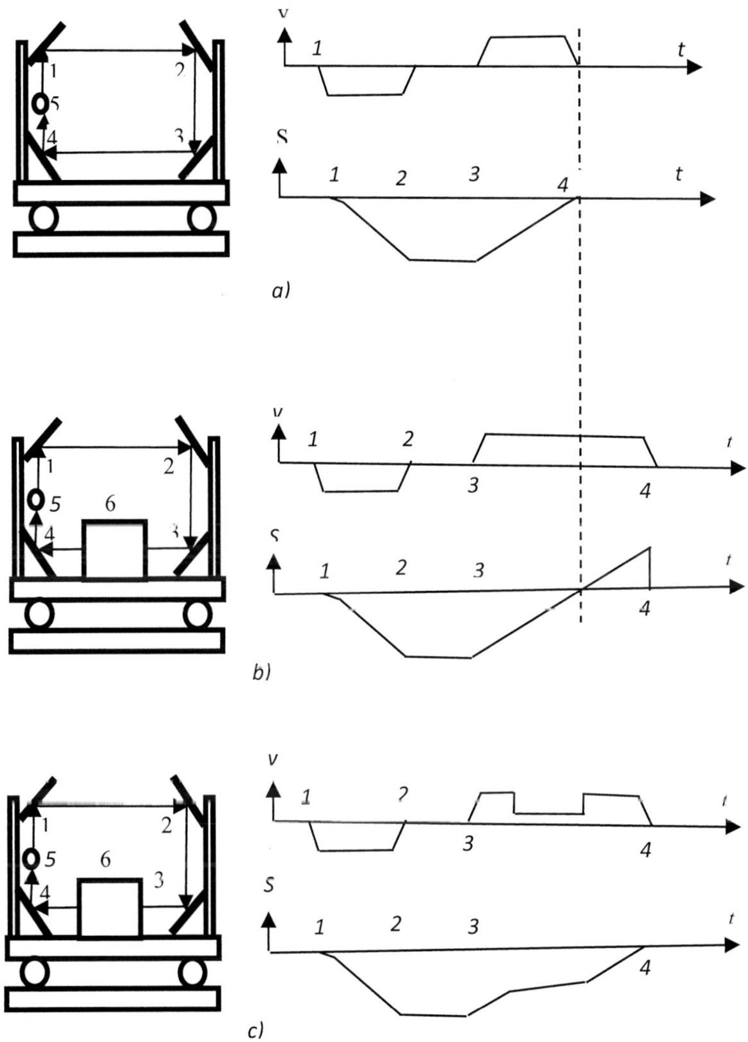

Fig. 2-1. Scheme of the modified Balazs thought experiment:

a) Dependences of the cart velocity and distance when a wave pulse is propagating in a homogeneous medium
b) Dependences of the cart velocity and distance when a wave pulse is propagating in an inhomogeneous medium where the pulse velocity in block 6 is decreased but the pulse does not interact with the medium

c) Dependences of the cart velocity and distance when a wave pulse is propagating in an inhomogeneous medium where the pulse velocity in block 6 is decreased but the pulse pushes the block in the direction of its propagation and decreases the velocity of the cart. A magnitude of the velocity decrease is required to determine.

It follows from the Eq. (2.4) that only a part of the pulse momentum passes into the mechanical movement of the cart, which depends on the ratio of the pulse velocities before the cart and in the cart.

It is this result that was obtained and interpreted by Balazs. The same interpretation was then repeated by Barnett and Loudon.

Our goal is to give a different interpretation of this experiment, so that it does not contradict other thought experiments, according to which the momentum of light in the block increases by v_1/v_2 times, but at the same time the block moves in accordance with Eq. (2.4).

Since the momentum of the pulse increases within the block, there is negative pressure P_F applied to the front face of the block and directed opposite the pulse propagation. The pressure P_F is given by

$$P_{Face} = W_1 - W_2 \tag{2.5}$$

Since the pressure is negative, the block must move in a negative direction, that is, opposite the direction of wave propagation. These contradictions can be eliminated if we take into account that Eq. (2.5) is true for a continuous wave. At the same time, Eq. (2.4) was obtained for a wave segment in the form of a wave pulse. The difference between a continuous wave and a pulse is that the pulse has edges where the energy density of the wave changes. It can be assumed that the change in energy density is accompanied by an appearance of additional pressure, the duration of which is equal to the duration of the negative pressure determined by Eq. (2.5) and is equal to the duration of the pulse.

If the speed of the wave in the block is v_1, then the momentum p_{Cart} in accordance with Eq. (2.4) is equal to zero. However, this does not mean that there is no pressure P_L produced by the leading edge of the pulse. In the absence of the block, the speed of the wave in the place where the block was located is equal to v_1 and the pulse can transfer momentum to the medium in the same way as it transfers momentum to a block in which the wave speed is different from v_1 and equal to v_2.

Taking into account that durations of the pulse T in mediums 1 and 2 are the same and the MFD of a wave is equal to its energy density W, the pressure P_{Cart} produced on the cart when the pulse is entering the block through its face in accordance with Eq. (2.4) can be presented as follows

$$P_{Cart} = W_1(1 - \frac{v_2}{v_1}) \tag{2.6}$$

A magnitude of new positive pressure P_{Add} produced by the leading edge of the pulse can be determined from the condition that a sum of this new pressure P_{Add} and the negative pressure Eq. (2.5) applied to the front face of the block is equal to the pressure - Eq. (2.6) - obtained from the Balazs thought experiment. Then this condition can be represented as follows

$$P_{Add} + P_{Face} = P_{Cart} \tag{2.7}$$

Passing from the energy density W to the energy flux density (wave intensity) S and taking into account that S inside and outside the block is the same, we have

$$S = W_1 v_1 = W_2 v_2 \tag{2.8}$$

Taking into account Eq. (2.5) and Eq. (2.6), we have from Eq. (2.7)

$$P_{Add} = W_1(1 - \frac{v_2}{v_1}) - W_1 + W_2 = -W_1\frac{v_2}{v_1} + W_2 = W_2(1 - \frac{v_2^2}{v_1^2}) \tag{2.9}$$

Passing from the pressure produced by the leading edge of the pulse to the density of forces arising in the medium under the action of the leading edge, we obtain

$$\boldsymbol{f}_{Add} = \frac{dP_{add}}{dl} = \frac{dW_2}{dl}\frac{v_2}{v_1}(\frac{v_1}{v_2} - \frac{v_2}{v_1})\boldsymbol{e} = \frac{dW_2}{dt}\frac{1}{v_1}(\frac{v_1}{v_2} - \frac{v_2}{v_1})\boldsymbol{e} \tag{2.10}$$

Taking into account Eq. (2.10), the force density within the block in medium 2 is given by

$$\boldsymbol{f}_{Add12} = \frac{dW_2}{dt}\frac{1}{v_1}(\frac{v_1}{v_2} - \frac{v_2}{v_1})\boldsymbol{e} = \frac{dS}{dt}\frac{1}{v_1}(\frac{v_1}{v_2} - \frac{v_2}{v_1})\frac{1}{v_2}\boldsymbol{e} = \frac{dS}{dt}(\frac{1}{v_2^2} - \frac{1}{v_1^2})\boldsymbol{e} \tag{2.11}$$

If the pulse passes from medium 2 to medium 3, where the propagation velocity is v_3, the additional force in medium 3 in accordance with Eq. (2.11) is given by

$$f_{Add23} = \frac{dS}{dt}\left(\frac{1}{v_3^2} - \frac{1}{v_2^2}\right)e \qquad (2.12)$$

If the pulse passes from medium 1 to medium 3, the additional force in medium 3 is given by

$$f_{Add13} = \frac{dS}{dt}\left(\frac{1}{v_3^2} - \frac{1}{v_1^2}\right)e \qquad (2.13)$$

Taking into account that the energy flux densities S in Eq. (2.12) and Eq. (2.13) are the same, we obtain the same result for the force in medium 3 if the pulse first passes from medium 1 to medium 2, and then from medium 2 to medium 3. Indeed,

$$f_{Add13} = f_{Add12} + f_{Add23} = \frac{dS}{dt}\left(\frac{1}{v_2^2} - \frac{1}{v_1^2}\right)e + \frac{dS}{dt}\left(\frac{1}{v_3^2} - \frac{1}{v_2^2}\right)e = \frac{dS}{dt}\left(\frac{1}{v_3^2} - \frac{1}{v_1^2}\right)e \qquad (2.14)$$

It can be concluded that when the pulse propagates in medium 2, it already, in accordance with Eq. (2.13), produces an additional force in this medium, and when it passes to medium 3, this additional force increases in accordance with Eq. (2.14).

From Eq. (2.14) it follows that the additional density of forces acting on the medium in which the wave propagates has the property of accumulation. When a wave passes from medium 1 into medium 3 through medium 2, the additional force density in medium 3 is determined by a sum of additional forces of mediums 1 and 2.

Thus, from the Balazs thought experiment, we can only determine the magnitude of additional forces that arise during the transition to a medium with a lower propagation velocity. However, we cannot determine the absolute magnitude of the forces, since additional forces from the previous medium are unknown.

First of all, it concerns the light wave. From the considered thought experiment, we can determine the additional forces that arise in the optical medium when light enters this medium from free space. However, there is no information about whether there are additional forces in free space. It is now generally accepted that such forces are zero. However, this assumption contradicts our main assumption that the increase in the momentum of any wave in time propagating in any medium is accompanied by the emergence of pressure acting on this medium.

The same result in a form of Eq. (2.14) is obtained on the assumption that the following relations take place

$$f_{Abs1=} = (\frac{dS}{dt}\frac{1}{v_1^2}) \qquad f_{Abs2} = (\frac{dS}{dt}\frac{1}{v_2^2}) \quad f_{Abs3} = (\frac{dS}{dt}\frac{1}{v_3^2}) \tag{2.15}$$

Here, f_{Abs} is the actual absolute value of the density force applied to the medium. The absolute density force in the next medium can be considered as the sum of the absolute density force in the previous medium and an additional density force equal to the difference between the absolute density force in the next medium and the density force in the previous medium. The absolute force density in medium 3 can then be presented as follows

$$f_{Abs3} = f_{Abs1} + f_{Add12} + f_{Add23} = f_{Abs1} + (f_{Abs2} - f_{Abs1}) + (f_{Abs3} - f_{Abs2}) = f_{Abs3} \tag{2.16}$$

By comparing Eq. (2.16) and Eq. (2.11) we can verify that Eq. (2.15) can be obtained from Eq. (2.11) by setting the second term to zero. Passing to the absolute forces in accordance with Eq. (2.15), we get that the additional forces Eq. (2.12) obtained from the Balazs thought experiment can be represented by the difference between the absolute forces in the medium with wave speed v_2 and in the medium with wave speed v_1.

Thus, the absolute density force can be presented as follows

$$f_{Abs} = (\frac{dS}{dt}\frac{1}{v_2^2}) = \frac{dW_2}{dt}\frac{1}{v_2} = \frac{dW_2}{dl} \tag{2.17}$$

Integrating Eq. (2.17) over the length of the pulse edge, we have

$$P_{Abs} = W_2 \tag{2.18}$$

To understand what Eq. (2.18) means, let us consider the propagation of a wave pulse of energy density W_2 at velocity v_2 in an optical medium along a z-axis, as shown in Fig. 2-2.

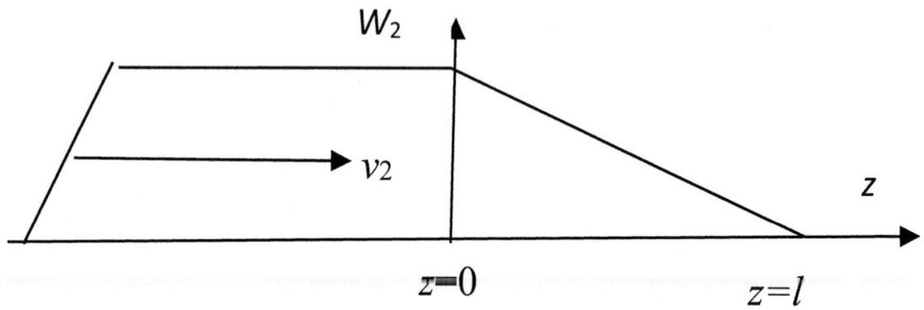

Fig. 2-2. Wave pulse of energy density W_2 propagating to the right along z-axis at speed v_2.

The pressure at the region $0 < z < l$ is equal to W_2 because the input MFD at cross-section z=0 is equal to W_2 and the output MFD at cross-section $z=l$ is equal to 0. Then, in accordance with Eq. (1.4), the pressure on this region $0<z<l$ is equal to W_2. The momentum that is transferred to this region of the medium is equal to MFDτ where $\tau = l/v_2$ is the duration of the leading pulse edge. The pressure is directed along the pulse propagation. Thus, the pressure P_{Abs} produced by the leading edge of the pulse in the region where it is located is equal to W_2.

This result is obtained on the assumption that W_2 and P_2 are equal to zero in the absence of the pulse. As for Eq. (2.4), it determines the increase in the pulse momentum in the block compared to the momentum of the same pulse before the block.

The above expressions can be interpreted as follows. The absolute density of forces is determined by the expressions in Eq. (2.15). When the pulse passes from the medium where the propagation velocity is v_1 to the medium where the propagation velocity is v_2, an additional force density arises equal to the difference between the indicated densities.

The same conclusion follows from Eq. (2.11). In this case, in accordance with our assumption, the first term is the absolute force density produced by the pulse in medium 2 and the second term is the absolute force density produced by the pulse in medium 1. This conclusion agrees with Eq. (2.18). Thus, the additional pressure in medium 2 is equal to the difference in absolute pressures in medium 2 and medium 1.

Another interpretation of the Balazs thought experiment

From the above considerations, it follows that the pulse completely transfers its momentum to the medium in which it propagates. When the pulse is propagating within the medium, the momentum of the medium is equal to the momentum of the pulse independent on the velocity of the pulse in the medium. Pressures applied to the medium in accordance with Eq. (2.5) and Eq. (2.18) are shown in Fig. 2-3.

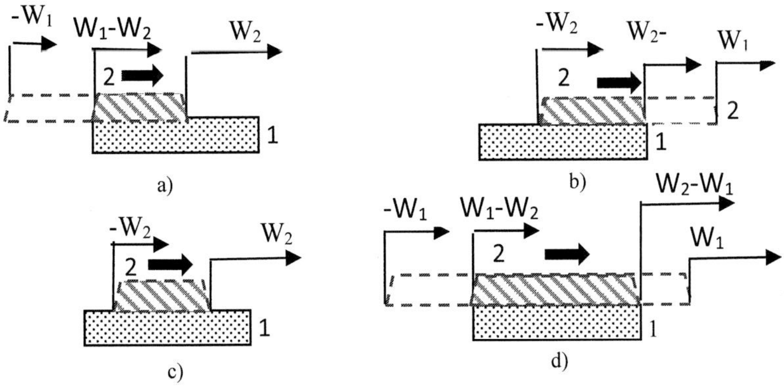

Fig. 2-3. Propagation of the light pulse 2 (dashed trapeze) in optical medium 1. The shaded portion of the pulse corresponds to the mechanical momentum that is propagating within the block.

Summing the pressures for each of the four cases in Fig. 2-3, we can make sure that the total pressure produced by the pulse on the medium is equal to zero and does not depend on the position of the pulse relative to mediums 1 and 2. The total pressure is always equal to zero.

In this case, the explanation of pulse propagation between reflectors, as shown in Fig. 2-1, is completely different from the explanation given earlier. When the pulse is reflected from reflector 1, two pressures arise. The first pressure $-W_1$ is negative and is applied to the reflector, that is, to the cart, because the reflector and the cart are an indivisible whole. This pressure arises in accordance with Newton's third law. The reflector produces the horizontal momentum of the pulse W_1 due to the force that acts from the side of the reflector on the pulse. In turn, the leading edge of the pulse produced in the medium pressure is W_1.

The pressure W_1 is positive and is applied to the medium in which the leading edge of the pulse begins to propagate horizontally. Ultimately, this pressure is also applied to the cart, since the medium and the cart are also an indivisible whole. According to Fig. 2-3a), this pressure is equal to W_1. Thus, two pressures act on the cart. One of them acts on the reflector, and the other on the medium in which the pulse propagates. The pressures are equal in magnitude and oppositely directed. The durations of action of these pressures are the same and equal to the pulse duration. As a result, the total pressure on the cart is zero and the cart remains stationary. The same is true for reflectors 2, 3, and 4.

Now, let us consider a situation in which block 6 in Fig. 2-1b) is inserted. The velocity of propagation within the block is v_2. In accordance with our conclusion, the total pressure applied to the medium with various velocities is equal to zero and does not depend on the position of the pulse relative to the boundary between these mediums. This means that the cart is stationary regardless of the speed v_2 in block 6. Thus, the cart is always stationary, whether there is block 6 or not.

Thus, our conclusions regarding the pressures and forces that act on the medium in which the pulse of an arbitrary wave propagates are fully consistent with the expectations of the Balazs thought experiment, according to which the cart should remain in its original state when the block 6 is inserted. It has been shown that this is indeed the case.

The above reasoning is applicable, as a special case, to an electromagnetic wave propagating in free space. It should be recognized that the same phenomena must take place in free space as in other mediums. The existence of the momentum of an electromagnetic wave in free space is not objectionable. However, the pressure on free space, which, in accordance with Newton's second law, inevitably arises when the momentum changes in time, according to modern ideas, cannot exist, although the basic law of mechanics is violated in this case.

This contradicts our findings. There are two ways out of this situation. Either we consider that our reasoning is incorrect and there are no forces in free space, or we consider that free space is affected by forces from the electromagnetic wave propagating in it. We have found a third way. We will consider both situations in the future and see what further consequences can be obtained using each of the two approaches.

The first approach assumes the conventional opinion that the pressure in free space is zero. In this case $P_{Abs}=P_{Add}$.

The second approach assumes that the force density and pressure in the free space are non-zero and are determined, as for any types of waves, by means of Eqs (2.15), (2.16), and (2.18).

We prefer the second option, since it directly follows from the generally accepted laws of physics and optics and is not based on any preliminary assumptions. Looking ahead, we can point out that when using the first approach, the formulas turn out to be more cumbersome, since they must take into account the features of the particular case associated with electromagnetic waves. It will be shown later that the formulas obtained on the basis of the second approach have a simpler form and contain fewer parameters, since there is no need to take into account the special case for a light wave in free space. In addition, the obtained formulas have a natural interpretation based on the known laws of electrical engineering. As is known, nature loves simplicity.

As for Fig. 2.1, it is valid only in one particular case when an electromagnetic wave propagating in free space violates Newton's second law, since the time-varying momentum does not create forces acting on free space. In the general case, the cart is always stationary, regardless of whether block 6 is present or absent. This is in full agreement with Newton's first law.

Forces produced by an arbitrary continuous travelling wave

The force density can also be produced by continuous waves. Actually, the MFD of the continuous wave is proportional to its energy density that is proportional to the square of the amplitude of the wave and, therefore, can be presented in a form of a sequence of pulses as follows

$$W = W_0 \sin^2(\omega t - \frac{2\pi}{\lambda} z) \tag{2.19}$$

As is seen from Eq. (2.19), the continuous wave is a sequence of pulses, the MFD of which is changed along the z-axis with space period $\lambda/2$. The velocity of this sequence is equal to $u=\lambda\omega/(2\pi)$. Since the MFD of the wave is changed with z, the wave produces forces applied to the region of the medium where the wave is propagating.

In the Balazs thought experiment, it was assumed that the averaged values of the energy of a wave pulse and a wave were considered. Expression (2.19) is also valid for such pulses, that is, Eq. (2.19) is valid not only for the considered average values, but also for the instantaneous values of the wave.

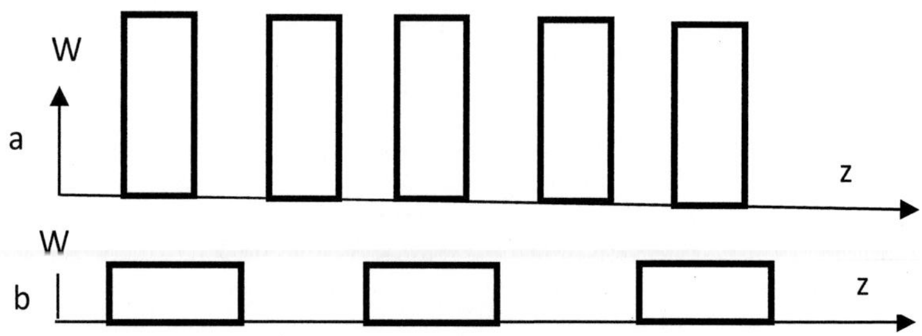

Fig. 2-4. Dependence of the energy density of a sequence of the energy density pulses on the length in medium.

a) Pulses are propagating in the block at velocity v_2.
b) Pulses are propagating before the block at velocity v_2.

For clarity, instead of pulses with a sinusoidal envelope, let us consider rectangular pulses. In Fig. 2-4, the upper graph shows the sequence of pulses propagating in the block at velocity v_2 and the energy density W_2. The lower graph shows the sequence of pulses propagating in medium 1 when the block is removed at speed v_1 and energy density W_1. It is assumed that $v_1>v_2$. In this case, the following relations are valid. Taking into account that durations of pulses in mediums 1 and 2 are identical and their MFDs are in inverse proportion with the wave velocity, the ratio of the momentums M_1 and M_2 of the pulses is given by

$$\frac{M_1}{M_2}=\frac{v_2}{v_1} \tag{2.20}$$

As can be seen from Fig. 2-4, the number of pulses N on a length equal to the length L of the block is inversely proportional to the wave velocity. In this case we have

$$\frac{N_1}{N_2}=\frac{v_2}{v_1} \tag{2.21}$$

The total momentum M_{L2} of pulses located in the block of length L is equal to the product M_2N_2. The total momentum M_{L1} of pulses located in the medium 1 of the same length L is equal to M_1N_1. Taking into account Eqs (2.20) and (2.21), we have

$$\frac{M_{L1}}{M_{L2}}=\frac{N_1M_1}{N_2M_2}=\frac{v_2}{v_1}\frac{v_2}{v_1}=(\frac{v_2}{v_1})^2 \qquad (2.22)$$

Thus, an increase in the momentum ΔM_L is equal to the difference M_{L2}-M_{L1}. Taking into account Eq. (2.22), we have

$$\Delta M_L=M_{L2}-M_{L1}=M_{L2}(1-\frac{M_{L1}}{M_{L2}})=M_{L2}[1-(\frac{v_2}{v_1})^2] \qquad (2.23)$$

Comparing Eq. (2.23) with Eq. (2.9), which was obtained from the agreement of results of the Balazs thought experiment with other generally accepted laws of physics and optics, one can make sure that the additional momentum imparted by the wave to the block in accordance with Eq. (2.23) corresponds to the additional pressure that should be produced in the block in order to provide the conditions for the conservation of momentum of an isolated system.

Thus, on the basis of simple considerations, we have obtained the same result as in the coordination of the Balazs thought experiment with other thought experiments based on generally accepted laws of physics and optics.

We can conclude that Eq. (2.18) is correct because it gives the absolute value of the force density, unlike Eq. (2.14) which gives the force density in medium 2 relative to the force density in medium 1.

It follows from general considerations that the force density produced by a wave in a medium does not depend on the medium from which the wave was introduced into the medium. In this case, the absolute pressure produced by the leading edge of the pulse is determined by Eq. (2.18) and not Eq. (2.9).

In this case we need to conclude that the momentum is related to the mechanical momentum due to the movement of the medium located between the edges of the pulse at the velocity of the wave in the medium. In the particular case of an electromagnetic wave, when $v_1=c$, $v_2=c/n$ and the wave outside the block propagates in free space, we have the following: the additional momentum transmitted by the wave to the block is equal to the difference between the absolute momentum of the block and the absolute momentum that the wave produces in free space in the region of L length.

This means that forces act on free space, which must be taken into account when coordinating the results of the Balazs thought experiment.

Capturing a medium by an electromagnetic wave

The energy flow density S of an electromagnetic wave is given by

$$S=[E\times H] \tag{2.24}$$

where E and H are amplitudes of the strength of electrical and magnetic fields, respectively (*Physical encyclopedia*, 1992).

Taking into account Eq. (2.8), we obtain from Eq. (2.17)

$$f=\frac{dW_2}{dt}\frac{1}{v_2}=\frac{dS}{dt}\frac{1}{v_2^2}=\frac{n^2}{c^2}\frac{dS}{dt}=\frac{n^2}{c^2}\frac{d[E\times H]}{dt} \tag{2.25}$$

In the particular case of free space where n=1, the forces acting on free space is given by

$$f=\frac{1}{c^2}\frac{d[E\times H]}{dt} \tag{2.26}$$

At present there is a generally accepted assumption that the free space is an emptiness and no force can be applied to free space. If so, Eq. (2.25) is not correct for the particular case of n=1 and should be corrected in such a way that f=0 at n=1. The correction is produced by subtracting Eq. (2.26) from Eq. (2.25). As a result, we have

$$f=\frac{n^2-1}{c^2}\frac{d[E\times H]}{dt} \tag{2.27}$$

Thus, there are two conflicting alternatives. If Newton's second law is correct, any wave must produce forces when its MFD changes in time. Since any other waves produce forces in the medium in which they propagate due to the change in their energy density over time, the electromagnetic wave must also obey the same laws. If there are no such forces in free space, then Newton's second law is violated. In accordance with this law, when the momentum of the electromagnetic wave changes, there must be forces responsible for this change. Our consideration is based completely and without any exceptions on Newton's laws, otherwise further consideration is meaningless. Out of respect for Newton, we cannot allow violations of Newton's laws.

Thus, a movement of a wave in a medium is accompanied by a movement of its momentum in a form of a mechanical momentum that is produced by forces connected with a movement of the wave energy in space. This wave is responsible for the movement of energy associated with the release of the medium from the state of rest in such a way that a mechanical momentum appears in the medium. This momentum exists between the leading and trailing edges of the pulse and travels along with the pulse at the speed of wave v_2. In this case, it is this mechanical momentum that is transferred to adjacent sections of the medium as the pulse travels to the right. A magnitude of this mechanical momentum is equal to the total momentum of the wave. By the way, the wave energy is also transferred by the medium where the wave is propagating.

Our notion about the process of wave propagation enables us to fully harmonize all known thought experiments based on generally accepted laws of physics and optics. If this notion is correct, then it becomes possible to fully explain the reason why the travelling continuous wave produces momentum in the medium in which it propagates. Any wave can be considered as the propagation of two types of energy in the form of a sequence of pulses. For a harmonic wave in a form $\sin(\omega t\text{-}2\pi z/\lambda)$, the energy envelope of these pulses is proportional to $\sin^2(2\pi z/\lambda)$, which period is $\lambda/2$. Each such pulse produces forces in the medium. The force is directed in the direction of the wave if the energy of the pulse increases with z and in the opposite direction if the energy of the pulse decreases.

As for to an electromagnetic wave, which is considered as a special case of a wave of an arbitrary type, it should be recognized that the momentum of the electromagnetic wave is completely determined by its mechanical momentum, which is produced by the electromagnetic wave in the medium where it is propagating.

Let us analyze, as an example, the capturing of air by a sound wave. In this case, there are experimental observations that an ultrasound wave captures the air where it is propagating (Mayer 1978), which confirm our theory. The propagation of a sound wave in air is accompanied by a displacement of the air in the direction of wave propagation. On the other hand, it must be taken into account that in each cross-section of the air medium, the energy density W of the sound wave changes periodically. As a result, a variable pressure in this-cross section acts on the medium. The pressure is directed along the wave propagation when W increases, and in the opposite direction when W decreases. Averaging this pressure over time, we find that the average pressure is zero.

However, this does not mean that air does not move through this cross-section. Indeed, under the action of positive pressure, the air moves in the direction of wave propagation. Under negative pressure, the air is shifted in the opposite direction, but it is another air that has taken the place of the air shifted in the positive direction. As a result, there is a movement of air in the direction of wave propagation.

For such movement of air, no energy consumption is required from the side of the sound wave. To start the movement of air, energy is required to give the air some speed. However, in order to bring the air to a motionless state, it is required to slow it down. In this case, the energy of the moving air is again transferred to the wave.

Perhaps it would be more understandable to consider this phenomenon in a moving coordinate system that moves at the speed of a sound wave. To simplify the consideration, let us assume that the energy density of the wave consists of positive rectangular pulses, the repetition rate of which is 2 times greater than the frequency of the wave. The duration of the pulses is equal to half the period of their repetition. In this case the pulse of the wave is motionless in the moving coordinate system but the air is moving in the opposite direction at the speed of the sound wave.

Positive pressure acts on the air in the cross-section in which the leading edge of the pulse is located. The pressure direction is positive and the pressure decreases the air velocity. In the cross-section in which the energy density decreases, the pressure direction is negative and the pressure resumes the air velocity. As a result, the velocity of the air between the edges of the pulse decreases. Returning to the motionless coordinate system, we see that the air velocity has become greater than zero due to the considered decrease of the negative air velocity between the edges of the pulses in the moving coordinate system.

Thus, we have 2 approaches for calculating the density of forces produced by a pulse as it propagates in a medium. In one approach, we can only determine the additional force density in medium 2 compared to the force density in medium 1. In another approach, we introduce absolute force density and assume that the additional force density derived from the Balazs thought experiment is equal to the difference between these absolute force densities for the two media.

An exception is the electromagnetic wave. According to modern concepts, the force density produced by an electromagnetic wave in free space is equal to zero, although the momentum of the wave changes in time, and in accordance with Newton's second law these changes should produce force densities. This exception to the general rule complicates the formulas for forces and pressures produced by electromagnetic waves and makes them difficult to interpret.

CHAPTER THREE

NEW TYPES OF FORCES PRODUCED BY TRAVELLING WAVES

In every branch of knowledge, the progress is proportional to the number of facts on which to build, and therefore to the facility of obtaining data.
J. Maxwell

What is common between the parameters of waves of various types

To better understand the properties of an electromagnetic wave, let's compare its properties with waves of other types, which propagate at a much lower speeds and were studied much earlier than electromagnetic waves were discovered. Waves of the most various physical natures are known. Their amplitudes are characterized by completely different parameters. However, for each wave there are general parameters, such as the speed of wave propagation, the energy and momentum densities, as well as the density of forces excited by the wave in the medium in which it propagates. However, the types of energy can be completely different. These are kinetic and elastic energy for an elastic wave, electric and magnetic energy for an electromagnetic wave, kinetic and potential energy in a gravitational field for a sea wave, etc. Therefore, the natures of the additional forces that arise when the energy density of the wave in space changes can be different, and are associated with the specifics of the wave energy.

As shown in the table below, for each type of wave there are 2 different types of forces associated with 2 different types of energy from which the wave consists. Although each type of wave is characterized by its own parameters, all waves have a lot in common, which makes it possible to compare their parameters. A similar situation occurs for such a physical notion as energy, which can have a variety of forms. As an illustration, Table 3-1 presents a comparison of the parameters of two completely different types of waves, which have completely different types of transferred energy and completely different parameters characterizing these waves.

Table 3-1. Comparison of parameters of elastic and electromagnetic waves.

For elastic wave			For electromagnetic wave	
Comments	Formulas		Comments	Formulas
Velocity of the displacement of medium in an elastic wave	v	1	E is the strength of an electrical field	E
P is the pressure of the compression	P	2	H is the strength of the magnetic field	H
Density of the medium	ρ	3	Permittivity of the medium	$\varepsilon_0\varepsilon$
Compressibility of the medium	β	4	Permeability of the medium	$\mu_0\mu$
Kinetic energy density	$W_k = \rho v^2 / 2$	5	Electrical energy density	$W_E = \varepsilon_0 \varepsilon E^2 / 2$
Elastic energy density	$W_P = \beta p^2 / 2$	6	Magnetic energy density	$W_H = \mu_0 \mu H^2 / 2$
W is the total energy density of the elastic wave	$W=W_K+W_P$	7	W is the total energy density of the light wave	$W=W_E+W_H$

Energy densities of kinetic and potential energies in the elastic wave are identical	$W = 2W_K = 2W_P = \rho v^2 = \beta p^2$	8	Energy densities of electrical and magnetic energies in the light wave are identical	$W = 2W_E = 2W_H = \varepsilon_0 \varepsilon E^2 = \mu_0 \mu H^2$
Velocity of propagation	$u = (\beta\rho)^{-1/2}$ $u_1 = (\beta_1\rho_1)^{-1/2}$ $u_2 = (\beta_2\rho_2)^{-1/2}$	9	U is velocity of light wave	$u = (\varepsilon_0 \varepsilon \mu_0 \mu)^{-1/2}$ $u_1 = (\varepsilon_0 \varepsilon \mu_0 \mu_1)^{-1/2}$ $u_2 = (\varepsilon_0 \varepsilon_2 \mu_0 \mu_2)^{-1/2}$
Elastic wave intensity	$S = vp = W(vp)^{-1/2} = Wu$	10	Light wave intensity	$S = [E \times H] = W(\varepsilon_0 \varepsilon \mu_0 \mu)^{-1/2} = Wu$
Equality of two types of energy	$W_K = W_P$	11	Equality of two types of energy	$W_E = W_H$
Resistance	$v / p = (\beta / \rho)^{1/2}$	12	Wave resistance	$E/H = (\mu_0 \mu / \varepsilon_0 \varepsilon)^{1/2}$
Relation between two amplitudes	$p = v(\rho / \beta)^{1/2}$	13	Relation between two amplitudes	$H = (\frac{\varepsilon_0 \varepsilon}{\mu_0 \mu})^{1/2} E$

P_{Add2} is the additional pressure produced by the front edge of the pulse when it penetrates from medium 1 where the wave velocity is u_1 to medium 2 where the wave velocity is u_2	$P_{Add2} = W_1(\frac{u_1^2}{u_2^2} - 1) = W_2 - W_1$	14	P_{Add2} is the additional pressure produced by the front edge of the pulse when it penetrates from medium 1 where the wave velocity is u_1 to medium 2 where the wave velocity is u_2	$P_{Add2} = W_1(\frac{u_1^2}{u_2^2} - 1) = W_2 - W_1$
W_2 is the energy density of the elastic wave in medium 2		15	W_2 is the energy density of the light wave in medium 2	

f_{Add} is the new type of the total density forces produced by the elastic wave	$f_{Add} = \frac{d(pv)}{dt}(\frac{1}{u_2^2} - \frac{1}{u_1^2})$ $(\frac{dp}{dt}v + \frac{dv}{dt}p)(\frac{1}{u_2^2} - \frac{1}{u_1^2}) =$ $f_{AddK} + f_{AddP}$ $f_{Add} = \frac{d(pv)}{dt}(\frac{1}{u_2^2} - \frac{1}{u_1^2})$ $(\frac{dp}{dt}v + \frac{dv}{dt}p)(\frac{1}{u_2^2} - \frac{1}{u_1^2}) =$ $f_{AddK} + f_{AddP}$ $f_{AddK} = (\frac{dp}{dt}v)(\frac{1}{u_2^2} - \frac{1}{u_1^2})$ $f_{AddK} = (p\frac{dv}{dt})(\frac{1}{u_2^2} - \frac{1}{u_1^2})$	16	f_{Add} is the new type of the total density forces produced by the light wave	$f_{Add} = \frac{d[E \times H]}{dt}(\frac{1}{u_2^2} - \frac{1}{u_1^2}) =$ $\{[\frac{dE}{dt} \times H] + [E \times \frac{dH}{dt}]\} \times$ $(\frac{1}{u_2^2} - \frac{1}{u_1^2}) =$ $f_{AddE} + f_{AddH}$ $f_{AddE} = [\frac{dE}{dt} \times H](\frac{1}{u_2^2} - \frac{1}{u_1^2}) =$ $[\frac{dE}{dt} \times H]\frac{1}{u_2^2}(1 - \frac{u_2^2}{u_1^2}) =$ $[\frac{dE}{dt} \times H]\varepsilon_0 \varepsilon \mu_0 \mu (1 - \frac{u_2^2}{u_1^2}) =$ $[\frac{dD}{dt} \times B](1 - \frac{u_2^2}{u^2})$ $f_H = [D \times \frac{dB}{dt}](1 - \frac{u_2^2}{u_1^2})$
	$f_{Abs} = \frac{d(pv)}{dt}\frac{1}{u_2^2}$ $(\frac{dp}{dt}v + \frac{dv}{dt}p)\frac{1}{u_2^2} =$ $f_{AbsK} + f_{AbsP}$	17		$f_{Abs} = \frac{d[E \times H]}{dt}\frac{1}{u_2^2} =$ $\{[\frac{dE}{dt} \times H] + [E \times \frac{dH}{dt}]\}\frac{1}{u_2^2} =$ $f_{AbsE} + f_{AbsH}$

f_{AbsK} is the new type of density force produced by the kinetic energy in the elastic wave	$f_{AbsK} = \frac{1}{u_2^2}(\frac{dp}{dt}v) =$ $(\beta_2\rho_2)(\frac{\beta_2}{\rho_2})^{-1/2}\frac{dv}{dt}v =$ $(\beta_2\rho_2)^{1/2}\frac{d}{dt}(\rho_2\frac{v^2}{2}) =$ $\frac{1}{u_2}\frac{dW_K}{dt} =$ $\frac{1}{u_2}\frac{dW_K}{dl}\frac{dl}{dt} =$ $\frac{dW_K}{dl} = \frac{1}{2}\frac{dW_{Total}}{dl}$	18	f_{AbsE} is the new type of density force produced by the kinetic energy in the elastic wave	$f_{AbsE} = [\frac{dD}{dt} \times B] =$ $\varepsilon_0\mu_0\varepsilon_2\mu_2[\frac{dE}{dt} \times H] =$ $\varepsilon_0\mu_0\varepsilon_2\mu_2[\frac{dE}{dt} \times (\frac{\varepsilon_0\varepsilon_2}{\mu_0\mu_2})^{1/2} E] =$ $(\varepsilon_0\mu_0\varepsilon_2\mu_2)^{1/2}\frac{d}{dt}(\varepsilon_0\varepsilon_2\frac{E^2}{2}) =$ $\frac{1}{u_2}\frac{dW_E}{dt} = \frac{1}{u_2}\frac{dW_E}{dl}\frac{dl}{dt} =$ $\frac{dW_E}{dl} = \frac{1}{2}\frac{dW_{Total}}{dl}$

f_{AbsP} is the new type of density force produced by the kinetic energy in the elastic wave	$f_{AbsP}=\frac{1}{u_2^2}(p\frac{dv}{dt})=$ $(\beta_2\rho_2)p(\frac{\beta_2}{\rho_2})^{1/2}\frac{dp}{dt}=$ $(\beta_2\rho_2)^{1/2}\frac{d}{dt}(\beta_2\frac{p^2}{2})=$ $\frac{1}{u_2}\frac{dW_P}{dt}=$ $\frac{1}{u_2}\frac{dW_P}{dl}\frac{dl}{dt}=$ $\frac{dW_P}{dl}=\frac{1}{2}\frac{dW_{Total}}{dl}$	19	f_{AbsH} is the new type of density force produced by the kinetic energy in the elastic wave	$f_{AbsH}=[D\times\frac{dB}{dt}]=$ $\varepsilon_0\mu_0\varepsilon_2\mu_2[E\times\frac{dH}{dt}]=$ $\varepsilon_0\mu_0\varepsilon_2\mu_2[(\frac{\varepsilon_0\varepsilon_2}{\mu_0\mu_2})^{-1/2}H\times\frac{dH}{dt}]=$ $(\varepsilon_0\mu_0\varepsilon_2\mu_2)^{1/2}\frac{d}{dt}(\mu_0\mu_2\frac{H^2}{2})=$ $\frac{1}{u_2}\frac{dW_H}{dt}=\frac{1}{u_2}\frac{dW_H}{dl}\frac{dl}{dt}=$ $\frac{dW_H}{dl}=\frac{1}{2}\frac{dW_{Total}}{dl}$
	$P_K=\int f_K dl=W_K$ $P_P=\int f_P dl=W_P$ $P_{Total}=W_K+W_P=W_{Total}$	20		$P_E=\int f_E dl=W_E$ $P_H=\int f_H dl=W_H$ $P_{Total}=W_E+W_H=W_{Total}$
	$W_K=W_P$ $f_K=f_P$ $P_K=P_P$	21		$W_E=W_H$ $f_E=f_H$ $P_E=P_H$

As can be seen from the table, the waves are characterized by completely different parameters. These are the electric and magnetic field strengths for electromagnetic waves E and H, respectively. These are the speed of particles of the medium v (not to be confused with the speed of wave propagation u) and the pressure p produced by the elastic wave in the medium (not to be confused with the pressure of the medium in which the wave propagates). In what follows, we will refer to such characteristics of an arbitrary wave as amplitudes.

Each wave is characterized by the propagation of two types of energy, as can be seen in lines 5 and 6. These are electrical and magnetic energy for an electromagnetic wave, and the kinetic energy and the energy of deformation of the medium for the elastic wave. The densities of these energies are proportional to the square of the corresponding amplitudes in lines 5 and 6. The densities of the two types of energies in each wave are the same. The factors at the squared amplitudes (ρ and β for an elastic wave as well as $\varepsilon_0\varepsilon$ and $\mu_0\mu$ for an electromagnetic wave) will be called the indexes.

As can be seen from lines 3, 4, and 9, the product of two indexes is equal to $1/u^2$, where u is the speed of wave propagation in the medium. Thus, knowing the dependences of the energy densities on the square of the amplitudes that characterize the given wave, it is possible to determine the propagation speed of this wave.

As can be seen from line 12, for each type of wave, the ratio of the two indicated amplitudes can be expressed as a ratio of the two mentioned indexes characterizing the wave.

As can be seen from line 10, for each type of a wave, the product of these amplitudes is equal to the intensity of the wave, that is, to the energy that the wave carries through the cross-section of a unit area per unit time. This intensity for each type of wave is equal to the product of the energy density of the wave and the speed of its propagation.

As can be seen from line 17, the additional pressure produced by the wave on the medium can be determined by the same formulas, regardless of the type of wave.

As can be seen from line 10, the wave intensity is determined by the same formulas if it is expressed in terms of the wave energy density. However, the energy density for waves of different types is expressed by completely different parameters, since the forms of the energy for different types of waves are completely different. At the same time, the energy of any type of wave consists of two types of energy, which ensure the propagation of the wave. These are the electrical and magnetic energy for electromagnetic waves, and the mechanical kinetic energy and the energy arising from compression (stretching) of the medium for elastic waves.

Each of the two types of energy is produced by two completely different types of additional forces that have different physical natures. The physical nature of the forces produced by an electromagnetic wave is discussed below.

The action of the above two additional types of force densities can ultimately be represented as the derivative of the density of the two types of energy with respect to the length along the propagation of the wave. Since these energies have different forms, the density forces that arise when these energies change in space have different physical origins. At the same time, the formulas that determine the magnitude of these forces are the same. This means that whatever the type of energy, the change in this energy in time and space leads to the emergence of the density forces that are described by the same formulas. However, the specification of each type of force when considering the process of a change of this type of energy in time or space leads to completely different formulas, depending on the types of energy under consideration. For example, an analysis of electromagnetic energy leads to formulas containing vectors E and H, which are connected with this type of energy. Specifying the energy of an elastic wave leads to formulas that contain completely different variables v and p. If formulas relating to the forces produced by an electromagnetic wave were previously known, albeit with errors, the formulas relating to the forces produced by an elastic wave are new, as far as we know.

It follows from the presented table that the physical natures of the additional forces are completely different, since these forces arise when completely different types of energy change in space. At the same time, the reasons for the emergence of these forces are the same and are associated with a change in energies in space, regardless of the form in which these energies are presented.

It is convenient to introduce common designations for parameters that characterized different waves for further consideration. These designations are presented in Table 3-2.

The parameters of an arbitrary wave can be characterized by three parameters. These are wave propagation velocity u, wave intensity S, and wave impedance R. The dimensions of the first two parameters are the same for all types of waves. The wave impedance determines the ratio of the two amplitudes that make up any wave. Since the dimensions of the amplitudes for different types of waves are different, the dimensions of the impedance are completely different and are determined by the specifics of the wave. From these three parameters, other more familiar parameters for any wave can be deduced, such as the amplitudes of the two waves A_1 and A_2, as well as the indexes B_1 and B_2, which determine the wave propagation speed u and its wave impedance R.

For the purpose of better understanding, we will compare the generalized parameters with the parameters of an electromagnetic wave. We present these comparisons in Table 3-2 without comments, since the above Table 3-1 already contains comments for the electromagnetic wave.

Table 3-2. Expression of the parameters of an arbitrary wave through three parameters (*S*, *u*, and *R*).

1	Intensity S	$S=A_1A_2$	$S=[E\times H]$
2	Wave speed u	$u=(B_1B_2)^{-1/2}$	$u=(\varepsilon_0\varepsilon\mu_0\mu)^{-1/2}$
3	Wave Resistance R	$R=A_1/A_2=(B_2/B_1)^{1/2}$	$R=E/H=(\varepsilon_0\varepsilon/\mu_0\mu)^{-1/2}$
4	Amplitude1 A_1	$A_1=(SR)^{1/2}$	$A_1=E$
5	Amplitude2 A_2	$A_2=(S/R)^{1/2}$	$A_2=H$
6	Index B_1	$B_1=u/R$	$B_1=\varepsilon_0\varepsilon$
7	Index B_2	$B_2=uR$	$B_2=\mu_0\mu$
8	Energy density1 W_1	$W_1=B_1A_1^2/2$	$W_1=\varepsilon_0\varepsilon E^2/2$
9	Energy density2 W_2	$W_2=B_2A_2^2/2$	$W_2=\mu_0\mu H^2/2$
10	Total energy density $W=S/u$	$W=S\sqrt{B_1B_2}$	$W=S(\varepsilon_0\varepsilon\mu_0\mu)^{1/2}$
11	Since $W_1 - W_2$	$(\frac{B_2}{B_1})^{1/2}=\frac{A_1}{A_2}$	$(\frac{\mu\mu}{\varepsilon\varepsilon})^{1/2}=\frac{E}{H}$
12	Total density force f_{Total}	$f_{Total}=$ $B_1B_2(\frac{dA_1}{dt}A_2+\frac{dA_2}{dt}A_1)=$ f_1+f_2	$f_{Total}=$ $(\varepsilon_0\varepsilon\mu_0\mu)([\frac{dE}{dt}\times H]+[E\times\frac{dH}{dt}])=$ f_1+f_2

13	Density force f_1	$f_1 = B_1 B_2 A_1 \frac{dA_2}{dt} =$ $B_1 B_2 \sqrt{SR} \frac{d\sqrt{S/R}}{dt} =$	$f_1 = [\frac{dD}{dt} \times B]$
14	Density force f_2	$f_2 = B_1 B_2 A_2 \frac{dA_1}{dt} =$ $B_1 B_2 \sqrt{S/R} \frac{d\sqrt{SR}}{dt} =$ $B_1 B_2 \sqrt{S} \frac{d\sqrt{S}}{dt}$	$f_2 = [D \times \frac{dB}{dt}]$
15	Density momentum M	$M = Wu = S/u^2 =$ $A_1 A_2 / u^2 = A_1 A_2 B_1 B_2$	$M = [E \times H]/u^2 =$ $(\varepsilon_0 \varepsilon \mu_0 \mu)[E \times H] =$ $[D \times B]$
16	$F_{total} =$ dM/dt	$\frac{dM}{dt} =$ $B_1 B_2 (\frac{dA_1}{dt} A_2 + \frac{dA_2}{dt} A_1)$ $B_1 B_2 (\frac{dA_1}{dt} (\frac{B_1}{B_2})^{1/2} A_1 +$ $\frac{dA_2}{dt} (\frac{B_2}{B_1})^{1/2} A_2) =$ $\frac{B_1}{2} \frac{dA_1^2}{dt} (B_1 B_2)^{1/2} +$ $\frac{B_2}{2} \frac{dA_2^2}{dt} (B_1 B_2)^{1/2} =$ $\frac{dW_1}{dt} \frac{1}{u} + \frac{dW_2}{dt} \frac{1}{u} =$ $\frac{1}{u} (\frac{dW_1}{dl} \frac{dl}{dt} + \frac{dW_1}{dl} \frac{dl}{dt})$ $= \frac{dW_1}{dl} + \frac{dW_2}{dl} = \frac{dW}{dl}$	$\frac{dM}{dt} =$ $\varepsilon_0 \varepsilon \mu_0 \mu ([\frac{dE}{dt} \times H] +$ $[E \times \frac{dH}{dt}]) =$ $\varepsilon_0 \varepsilon \mu_0 \mu (\frac{dE}{dt} (\frac{\varepsilon_0 \varepsilon}{\mu_0 \mu})^{1/}$ $+ H (\frac{\mu_0 \mu}{\varepsilon_0 \varepsilon})^{1/2} \frac{dH}{dt})$ $\frac{\varepsilon_0 \varepsilon}{2} \frac{dE^2}{dt} (\varepsilon_0 \varepsilon \mu_0 \mu)^{1/}$ $\frac{\mu_0 \mu}{2} \frac{dH^2}{dt} (\varepsilon_0 \varepsilon \mu_0 \mu)$ $\frac{dW_E}{dt} \frac{1}{u} + \frac{dW_H}{dt} \frac{1}{u} =$ $\frac{dW_E}{dl} \frac{dl}{dt} \frac{1}{u} + \frac{dW_H}{dt} \frac{d}{d}$ $\frac{dW_E}{dl} + \frac{dW_H}{dl} = \frac{dW}{dl}$
17	P_{Total}	W	W

Thus, the mass wave can be characterized by the parameters defined in the first three rows of the tables. Usually, wave speed and wave impedance are given in textbooks. The third parameter is the intensity S of the wave. Lines 4 and 5 describe the amplitudes A_1 and A_2 of the two waves that are associated with any type of wave. These amplitudes characterize two phenomena occurring during the propagation of arbitrary waves. Lines 6 and 7 contain indices B_1 and B_2, which determine the energy densities of the two waves. Lines 8-10 define the energy densities of the two waves and the total energy density of the waves. Bearing in mind the previous lines, we can verify that the energy densities W_1 and W_2 of each of the two waves are the same. Lines 12-14 define the force densities produced by each of the two waves and the total force density. Line 15 defines the total momentum density of the wave, and line 16 defines the total force density. It can be seen that this force density does not depend on the specifics of the wave, but is completely determined by the energy density of the wave. Line 17 is obtained by integrating the expression in line 16 over the length where an increase in the pulse density occurs. The total pressure is equal to the maximum energy density in the wave, which agrees with the previous considerations and Eq. (2.18).

Since two energies in any wave have different forms, the forces that arise when these energies are changed in space have different physical origins. At the same time, the formulas that determine the magnitude of these forces are the same for different waves but the physical origin of these forces for each particular case can be perfectly different since they originate from different types of energy.

All of the above applies to the electromagnetic wave. In this case, the two types of different energy are well known. They are the energy of electric and magnetic fields, respectively. When comparing these energies with the two energies W_1 and W_2 in a wave of an arbitrary type with amplitudes A_1 and A_2, it is impossible to determine which of the two energies of the electromagnetic wave corresponds to W_1 or W_2, since the formulas for the two types of energies are identical in form, but different in the physical meaning.

If we consider an electromagnetic wave as a special case of a wave of an arbitrary type, then it should be recognized that the total force acting on the medium is determined by line 16. It follows from this that such a force exists in free space, where $\varepsilon=1$ and $\mu=1$. If we assume that in free space an electromagnetic wave does not produce a force, then it should be recognized that in this case Newton's second law does not work, since in this case the electromagnetic wave has momentum in free space that is changed in time, but when this momentum is changed there are no forces that should be a cause of its change.

New types of forces produced by waves of various types

New type of forces in sound waves in gases

It is known that unlike an elastic wave the velocity of a sound wave in gases is given by

$$u^2 = \gamma \frac{P_{gas}}{\rho} \tag{3.1}$$

where P_{gas} is the gas pressure where the sound wave is propagating, and the dimensionless index γ is given by

$$\gamma = c_P / c_V \tag{3.2}$$

Here, c_P and c_V are heat capacities at constant pressure and constant volume, respectively.

The densities of the kinetic and potential energies are given by

$$W_K = \rho \frac{v^2}{2} \tag{3.3}$$

In this case, we see in Table 3-2 that $B_1 = \rho$. Then, from line 9 of the table and Eq. (3.1) we have

$$B_2 = \frac{1}{\gamma P_{gas}} \tag{3.4}$$

In this case, the potential energy density is given by

$$W_P = \frac{1}{\gamma P_{gas}} \frac{p^2}{2} \tag{3.5}$$

The parameter R for a sound wave depends on the medium where the wave is propagating. For example, R=340 kg $s^{-1}m^{-2}$ for the air at normal condition.

There are two new identical density forces due to changes to the kinetic and potential energies in space when the wave is propagating at velocity u. Since $p=(SR)^{1/2}$, $v=(S/R)^{1/2}$ and $u^{-2}=AB$, we see from Table 3-2, lines 13 and 14, that

$$f_K = \frac{\sqrt{SR}}{u^2}\frac{d\sqrt{S/R}}{dt} = \frac{p}{u^2}\frac{dv}{dt} \tag{3.6}$$

$$f_P = \frac{\sqrt{S/R}}{u^2}\frac{d\sqrt{SR}}{dt} = \frac{v}{u^2}\frac{dp}{dt} \tag{3.7}$$

A physical origin of the force in Eq. (3.6) is the following. The term dv/dt describes an acceleration and the dimension of p/u^2 is the mass density. A product of mass density and acceleration is the density force. The dimensions of p/u^2 and W/u^2 are identical. Following the known expression $m=E/c^2$, we can assume that the product pv is an analog of the Newton momentum where mass is presented in an energetic form. Then, Eq. (3.6) describes the force when a change of the momentum is caused by a change of the velocity and Eq. (3.7) describes the force arising due to a change of the momentum caused by a change of the mass presented in the energetic form.

New types of forces in stretched string waves

The parameters of a traveling wave in a stretched string are determined as follows.

The wave speed is determined by the expression

$$u = 1/\sqrt{B_1 B_2} = \sqrt{\frac{T}{\rho}} \tag{3.8}$$

Here ρ is the linear density of the string [kg/m] and T is string tension [N=kg m s^{-2}].

The densities of the kinetic energies and potential energies per unit of length are given by

$$W_K = \rho \frac{v^2}{2} \tag{3.9}$$

$$W_P = \frac{1}{T} \frac{f^2}{2} \tag{3.10}$$

where f is the force applied to the velocity of the string in transverse direction when the string deviates from equilibrium.

There are two new identical density forces due to a change of the kinetic and potential energies in space when the wave is propagating at velocity u. These forces are given by

$$f_K = \frac{dW_K}{dl} = \frac{d}{dl}(\rho \frac{v^2}{2}) \tag{3.11}$$

$$f_P = \frac{dW_P}{dl} = \frac{d}{dl}(\frac{1}{T} \frac{f^2}{2}) \tag{3.12}$$

Parameter R is given by

$$R = \sqrt{B_2 / B_1} = \sqrt{1/(T\rho)} \tag{3.13}$$

There are two new identical density forces due to a change of the kinetic and potential energies in space when the wave is propagating at velocity u. Since $p=(SR)^{1/2}$, $v=(S/R)^{1/2}$ and $u^{-2}=AB$, we see from Table 3-2, lines 13 and 14, that

$$f_K = \frac{\sqrt{SR}}{u^2} \frac{d\sqrt{S/R}}{dt} = \frac{f}{u^2} \frac{dv}{dt} \tag{3.14}$$

$$f_P = \frac{\sqrt{S/R}}{u^2} \frac{d\sqrt{SR}}{dt} = \frac{v}{u^2} \frac{df}{dt} \tag{3.15}$$

New types of forces in torsional waves in a thin-walled pipe

The velocity of the wave is given by

$$u = \sqrt{\frac{B_2}{B_1}} = \sqrt{\frac{G}{\rho}} \tag{3.16}$$

Here

$$G = \frac{E(1+\nu)}{2} \tag{3.17}$$

where E is Young's modulus and ν is Poisson's ratio.

The densities of the kinetic and torsion energies are given by

$$W_K = \rho\frac{v^2}{2} \tag{3.18}$$

$$W_P = \frac{p^2}{2G} \tag{3.19}$$

There are two new, identical density forces due to the change of the kinetic and potential energy u in space when the wave is propagating at velocity u. These forces are given by

$$f_K = \frac{d}{dl}(\rho\frac{v^2}{2}) \tag{3.20}$$

$$f_P = \frac{d}{dl}(\frac{1}{G}\frac{p^2}{2}) \tag{3.21}$$

New types of forces in gravity waves in shallow water

As is known, the velocity of the wave is given by

$$u = \sqrt{\frac{\rho g H}{\rho}} = \sqrt{gH} \tag{3.22}$$

Here H is the maximal water height, g is the free fall acceleration, and p is the overpressure produced by the wave.

The densities of the kinetic and potential energies are given by

$$W_K = \rho \frac{v^2}{2} \tag{3.23}$$

In this case, we see in Table 3-2 that $B_1=\rho$. Then from, line 6 of Table 2, we have

$$B_2 = \frac{1}{\rho g H} \tag{3.24}$$

Then the potential energy density is given by

$$W_P = \frac{1}{\rho g H} \frac{p^2}{2} \tag{3.25}$$

There are two new, identical density forces due to the change of the kinetic and potential energy in space when the wave is propagating at velocity u. These forces are given by

$$f_K = \frac{d}{dl}(\rho \frac{v^2}{2}) \tag{3.26}$$

$$f_P = \frac{d}{dl}(\frac{1}{\rho g H} \frac{p^2}{2}) \tag{3.27}$$

As follows from Eq. (3.27), the force f_P contains as a multiplier the acceleration of free fall g. Therefore, the type of force f_P differs from all other previously considered forces.

New types of forces produced by an electromagnetic wave

We have shown that the momentum flux density of light in an optical medium is equal to the energy density of the light. Then, in accordance with Newton's second law, one can calculate the forces acting on light, which are responsible for changing its momentum. These forces act from the side of the optical medium. According to Newton's third law, forces acting on the optical medium are equal in magnitude and oppositely directed.

The momentum of a light pulse in an optical medium increases by n times compared to that of the same light pulse in free space, where n is the ratio of the speed of light in free space to the speed of light in an optical medium. In general, n is determined by the expression

$$n=\sqrt{\varepsilon\mu} \tag{3.28}$$

Then the momentum of light pulse **p** in an optical medium is given by

$$\mathbf{p}=\mathbf{e}\frac{\mathscr{E}}{c}\sqrt{\varepsilon\mu} \tag{3.29}$$

where $\mathscr{E}$ is the energy of light and **e** is the unit vector directed along the direction of propagation of light.

The force applied to the light in accordance with Newton's second law is given by

$$\mathbf{F}=\frac{d\mathbf{p}}{dt}=\frac{\mathscr{E}}{c}\frac{d(\mathbf{e}\sqrt{\varepsilon\mu})}{dt}+\frac{\mathbf{e}\sqrt{\varepsilon\mu}}{c}\frac{d\mathscr{E}}{dt} \tag{3.30}$$

Given the force density f, the momentum density $\boldsymbol{\pi}$ and the energy density W in the optical medium, Eq. (3.30) can be presented as follows

$$\mathbf{f}=\frac{d\pi}{dt}=\frac{W}{c}\frac{d(\mathbf{e}\sqrt{\varepsilon\mu})}{dt}+\frac{\mathbf{e}\sqrt{\varepsilon\mu}}{c}\frac{dW}{dt} \tag{3.31}$$

Thus, the force acting on the optical medium consists of two components. One component is related to the fact that the momentum of light changes with a change in the reflective index when light propagates in an inhomogeneous optical medium. It is assumed that the energy of the light pulse remains unchanged. Another component is related to the change in the momentum of light with a change in its energy. It is assumed that light propagates in a homogeneous medium. Let's consider each component separately.

Forces produced by a light pulse in an inhomogeneous optical medium

Let us first consider a special case when the energy of light in free space W_0, which is introduced into the optical medium, does not change in time. This is the most typical case when a continuous light wave of constant intensity is introduced into the optical medium, and W_0 represents the average density of this wave. Since the energy density of the light W in an optical medium is inversely proportional to its speed, then, in accordance with Eq. (3.28), we have

$$W = W_0\sqrt{\varepsilon\mu} \tag{3.32}$$

In this case, the force density determined by the first term in Eq. (3.31) can be presented as follows

$$\mathbf{f} = \frac{W_0\sqrt{\varepsilon\mu}}{c}\frac{d(\mathbf{e}\sqrt{\varepsilon\mu})}{dl}\frac{dl}{dt} = \frac{W_0\sqrt{\varepsilon\mu}}{c}\frac{d(\mathbf{e}\sqrt{\varepsilon\mu})}{dl}\frac{c}{\sqrt{\varepsilon\mu}} = W_0\frac{d(\mathbf{e}\sqrt{\varepsilon\mu})}{dl} \tag{3.33}$$

where dl is an element of the distance along the trajectory.

It is known from geometric optics that the trajectory of light propagation in an inhomogeneous optical medium is given by (Landsberg 1976)

$$grad(n) = \frac{d}{dl}(n\mathbf{e}) \tag{3.34}$$

Here, n is the reflective index of the optical medium, equal to the ratio of the speed of light in free space to the speed of light in the optical medium.

Then from Eqs (3.28), (3.33), and (3.34) we have

$$\mathbf{f} = W_0 grad(\sqrt{\varepsilon\mu}) \tag{3.35}$$

Thus, the force per unit volume acting on light does not depend on the direction of its propagation and is always parallel to the gradient of the generalized reflective index.

Since the energy densities of the electric and magnetic fields in an electromagnetic wave are the same, Eq. (3.35) can be represented as follows

$$\mathbf{f} = \frac{W_0}{2}\frac{\mu grad(\varepsilon) + \varepsilon grad(\mu)}{\sqrt{\varepsilon\mu}} = \frac{W_0}{2}\sqrt{\varepsilon\mu}(\frac{grad(\mu)}{\mu} + \frac{grad(\varepsilon)}{\varepsilon}) = \frac{W_E + W_H}{2}(\frac{grad(\mu)}{\mu} + \frac{grad(\varepsilon)}{\varepsilon}) = W_E\frac{grad(\varepsilon)}{\varepsilon} + W_H\frac{grad(\mu)}{\mu} \quad (3.36)$$

where W_E and W_H are time-averaged energy densities of the electric and magnetic fields in the optical medium, respectively.

These densities are given by

$$W_E = \varepsilon_0\varepsilon\frac{E^2}{4}, \quad (3.37)$$

$$W_H = \mu_0\mu\frac{H^2}{4}. \quad (3.38)$$

where E and H are amplitudes of the electrical and magnetic fields. Then, Eq. (3.36) can be presented as follows

$$\mathbf{f} = \varepsilon_0 grad(\varepsilon)\frac{E^2}{4} + \mu_0 grad(\mu)\frac{H^2}{4} \quad (3.39)$$

Bearing in mind Newton's third law and passing from forces acting on light (Eq. (3.39)) to forces acting on an optical medium, we obtain that the following forces act on the optical medium

$$\mathbf{f} = -\varepsilon_0 grad(\varepsilon)\frac{E^2}{4} - \mu_0 grad(\mu)\frac{H^2}{4} \quad (3.40)$$

The same result can be obtained in another way. It is known from electrostatics and magneto statics that in static electric and magnetic fields the forces acting on a dielectric are determined as follows (*Physical encyclopedia*, 1992)

$$\mathbf{f} = -\varepsilon_0 grad(\varepsilon)\frac{E_S^2}{2} - \mu_0 grad(\mu)\frac{H_S^2}{2} \quad (3.41)$$

where E_S and H_S are electric and magnetic field strengths, respectively.

Averaging Eq. (3.41) over time, we obtain Eq. (3.40).

Thus, from the assumption that the momentum of light in an optical medium increases by a factor $\sqrt{\varepsilon\mu}$, the same expression - Eq. (3.40) - follows as from the laws of electrostatics and magneto statics used to calculate forces in alternating electric and magnetic fields.

Carrying out transformations in the reverse order from Eq. (3.41) to Eq. (3.32), we see that the increase in the momentum of light by a factor $\sqrt{\varepsilon\mu}$ follows from the equations of statics, which were known even before it became known that light was an electromagnetic wave. This circumstance is another argument for the momentum of light in an optical medium increasing by a factor $\sqrt{\varepsilon\mu}$.

It follows from the above consideration that the optically induced forces acting on an inhomogeneous optical medium are known and can be determined from electrostatics and electrodynamics. In accordance with Eq. (3.35), these forces are proportional to the product of the average energy density of light and the gradient of the generalized reflective index. The direction of these forces does not depend on the direction of wave propagation and is determined by the direction of the generalized reflective index. This is a non-trivial result, since, in accordance with the currently popular approach based on the use of the Lorentz law, these forces depend on the direction of the strengths of alternating electric and magnetic fields. A comparison of these two approaches is given below.

Forces produced in a homogeneous optical medium

Let us now consider the force density that emerges when the energy density changes over time. In accordance with Eq. (3.31) we have

$$\mathbf{f} = \frac{\mathbf{e}\sqrt{\varepsilon\mu}}{c}\frac{dW}{dt} = \frac{\mathbf{e}\sqrt{\varepsilon\mu}}{c}\frac{d}{dt}\left(\frac{S\sqrt{\varepsilon\mu}}{c}\right) = \mathbf{e}\frac{\varepsilon\mu}{c^2}\frac{dS}{dt} \tag{3.42}$$

Here S[Watt/m^2] is the energy flux density given by

$$\mathbf{S} = \mathbf{e}Wv = \mathbf{e}W\left(\frac{c}{\sqrt{\varepsilon\mu}}\right) \tag{3.43}$$

It is known that the intensity of an electromagnetic wave is given by (Landau L. D., Lifshits E. M., and Pitaevski L. P. 1976)

$$\boldsymbol{S} = [\boldsymbol{E} \times \boldsymbol{H}] \tag{3.44}$$

Bearing in mind that $c^{-2}=\varepsilon_0\mu_0$ and taking into account Eq. (3.42) and Eq. (3.44), we obtain

$$\boldsymbol{f} = \frac{\varepsilon\mu}{c^2}\frac{d[\boldsymbol{E}\times\boldsymbol{H}]}{dt} = \varepsilon_0\mu_0\varepsilon\mu\frac{d[\boldsymbol{E}\times\boldsymbol{H}]}{dt} = \frac{d[\boldsymbol{D}\times\boldsymbol{B}]}{dt} = [\frac{d\boldsymbol{D}}{dt}\times\boldsymbol{B}]+[\boldsymbol{D}\times\frac{d\boldsymbol{B}}{dt}] \tag{3.45}$$

Here $\boldsymbol{D} = \varepsilon_0\varepsilon\boldsymbol{E}$ and $\boldsymbol{B} = \mu_0\mu\boldsymbol{H}$ are displacement and induction, respectively.

Since Eq. (3.45) follows from generally accepted laws, such as Newton's laws, Maxwell's equations, Snell's law, Doppler's law, etc., it can be argued that it is a consequence of these laws that is derived from these laws without any additional assumptions about the nature of the forces in question.

Taking into account that the resistance of an electromagnetic wave is given by

$$R = \frac{E}{H} = \sqrt{\frac{\mu_0\mu}{\varepsilon_0\varepsilon}} \tag{3.46}$$

and the velocity of the electromagnetic wave is given by

$$u = \frac{E}{H} = (\varepsilon_0\varepsilon\mu_0\mu)^{-1/2} \tag{3.47}$$

as well as Eq. (3.37) and Eq. (3.38), we can present the first and seconds terms in Eq. (3.45) as follows

$$\begin{gathered} f_E = [\frac{dD}{dt}\times B] = \varepsilon_0\mu_0\varepsilon\mu[\frac{dE}{dt}\times H] = \varepsilon_0\mu_0\varepsilon\mu[\frac{dE}{dt}\times(\frac{\varepsilon_0\varepsilon}{\mu_0\mu})^{1/2}E] = \\ (\varepsilon_0\mu_0\varepsilon\mu)^{1/2}\frac{d}{dt}(\varepsilon_0\varepsilon\frac{E^2}{2}) = \frac{1}{u}\frac{dW_E}{dt} = \frac{1}{u}\frac{dW_E}{dl}\frac{dl}{dt} = \frac{dW_E}{dl} \end{gathered} \tag{3.48}$$

$$\begin{gathered} f_H = [D\times\frac{dB}{dt}] = \varepsilon_0\mu_0\varepsilon\mu[E\times\frac{dH}{dt}] = \varepsilon_0\mu_0\varepsilon\mu[(\frac{\varepsilon_0\varepsilon}{\mu_0\mu})^{-1/2}[H\times\frac{dH}{dt}] = \\ (\varepsilon_0\mu_0\varepsilon\mu)^{1/2}\frac{d}{dt}(\mu_0\mu\frac{H^2}{2}) = \frac{1}{u}\frac{dW_H}{dt} = \frac{1}{u}\frac{dW_H}{dl}\frac{dl}{dt} = \frac{dW_H}{dl} \end{gathered} \tag{3.49}$$

As is seen from Eq. (3.48) and Eq. (3.49), forces produced by the electromagnetic wave have a simple physical interpretation. These are forces that arise due to changes in the energy density in space due to the movement of the energy in space. One of these forces is responsible for changing the energy density connected with the electric field and the other is responsible for changing the energy density connected with the magnetic field. Since these densities are equal, the forces caused by the change in these densities are also equal. The resultant force is equal to the sum of these forces or twice each of them.

In accordance with Eq. (3.45), the density of forces in free space is nonzero and is given by

$$\boldsymbol{f}=\frac{1}{c^2}\frac{d[\boldsymbol{E}\times\boldsymbol{H}]}{dt} \tag{3.50}$$

Under this assumption, the validity of which we do not know, the force density in free space must be equal to zero, although we have shown that there are forces connected with a change of the energy in space or in time. We can satisfy this assumption if we subtract Eq. (3.50) from Eq. (3.45). In this case, we have

$$\begin{aligned}\boldsymbol{f}&=\frac{\varepsilon\mu-1}{c^2}\frac{d[\boldsymbol{E}\times\boldsymbol{H}]}{dt}=\frac{\varepsilon\mu(1-1/\varepsilon\mu)}{c^2}\frac{d[\boldsymbol{E}\times\boldsymbol{H}]}{dt}=\\ &\varepsilon_0\mu_0\varepsilon\mu(1-1/\varepsilon\mu)\frac{d[\boldsymbol{E}\times\boldsymbol{H}]}{dt}=(1-1/\varepsilon\mu)\frac{d[\boldsymbol{D}\times\boldsymbol{B}]}{dt}\end{aligned} \tag{3.51}$$

Eq. (3.51) coincides with the well-known expression that determines the Abraham forces in an optical medium, the existence of which is currently not generally accepted. Mentions of the Abraham forces can be found in reference of books and in references in some articles (*Physical encyclopedia*, 1992; Brevik 2009, 219301).

A project for an experimental facility for the experimental detection of these forces has been proposed (Brevik 2009, 219301).

Thus, there are two conflicting alternatives. If Newton's second law is true, then an electromagnetic wave should be subject by forces in free space when its momentum changes. These forces are applied to electromagnetic fields in the void. On the other hand, if such forces are absent, then Newton's second law is violated in free space, since, in accordance with this law, when the momentum of light changes, there must be forces responsible for this change. Out of respect for Newton, we cannot allow Newton's laws to be violated. In what follows, we will consider only the first option. The second option differs from the first only by a factor of (1-1/$\varepsilon\mu$).

Thus, in a general case the density of optically induced forces in accordance with Eqs (3.41) and (3.45) can be presented as follows

$$f = -grad(\varepsilon)\varepsilon_0 \frac{E^2}{2} - grad(\mu)\mu_0 \frac{H^2}{2} + \frac{d[\boldsymbol{D} \times \boldsymbol{B}]}{dt} \tag{3.52}$$

Since Eq. (3.52) follows from generally accepted laws, such as Newton's laws, Maxwell's equations, Snell's law, Doppler's law, etc., it can be argued that it is a consequence of these laws. This consequence is deduced from these laws without any additional assumptions about the nature of the forces in question.

Moreover, as is shown below, the Abraham forces produced by an electromagnetic wave is a particular case of waves of an arbitrary physical nature, in particular, of sound waves. There is an experimental confirmation of the existence of the Abraham forces produced by sound waves (Mayer 1978).

Comparison with the alternative approach based on the Lorentz law

Unlike forces produced by waves of an arbitrary type, forces produced by alternating electric and magnetic fields have been studied for more than a century. Formulas for calculating the forces acting from alternating electric and magnetic fields in the general case were first proposed by A. Einstein and J. Laub in 1908 (Einstein and Laub 1908). In 1973, formulas were proposed for the particular case when alternating magnetic and electric fields are produced by an electromagnetic wave (Gordon 1973, 14). One of the latest formulas for calculating optically induced forces was proposed in 2013, that is, more than a hundred years after the beginning of research in this direction (Mansiripur, Zakharian and Wright 2013, 023826). There are currently hundreds of papers that calculate the forces produced by electromagnetic waves based on the approach using Lorentz forces. Moreover, some authors specifically indicate that they used such an approach, implicitly making it clear that they did not trust the approach based on the analysis of a change in the momentum of light.

A comparison of various currently-known formulas for calculating forces in alternating electric and magnetic fields using the Lorentz law was given in Mansiripur, Zakharian and Wright (2013, 023826). However, there is no conclusion about which of the known formulas is correct or whether there is even one among them. It concludes that "further analysis is required to identify the correct formulas that could accurately predict the measured parameters." Thus, at present it is impossible to obtain reliable results with the approach based on the use of Lorentz forces.

Knowing the reliable expressions for the particular case when forces are produced in a homogeneous medium in the field of an electromagnetic wave, it is possible to analyze the known expressions for optically induced forces in the field of an electromagnetic wave and compare them with those based on the Lorentz forces approach.

However, first, one should present the formulas for calculating the density of forces in an optical medium based on the Lorentz laws, as proposed in the most authoritative 1908 publication by A. Einstein and J. Laub.

The forces $\mathbf{f}_1$, $\mathbf{f}_2$, $\mathbf{f}_3$, and $\mathbf{f}_4$ arising in a medium under the action of alternate electric and magnetic fields changed in time were presented as follows.

$$\begin{aligned} f_1 &= (\boldsymbol{P}\nabla)\boldsymbol{E} \\ f_2 &= (\boldsymbol{M}\nabla)\boldsymbol{H} \\ \mathbf{f}_3 &= [\frac{d\mathbf{P}}{dt} \times \mu_0 \boldsymbol{H}] \\ \mathbf{f}_4 &= [\varepsilon_0 \mathbf{E} \times \frac{d\mathbf{M}}{dt}] \end{aligned} \tag{3.53}$$

The forces $\mathbf{f}_3$ and $\mathbf{f}_4$ are the forces arising in an inhomogeneous medium under the action of alternate electric and magnetic fields. Parameters **P** and **M** are the polarization and magnetization vectors of the medium given by

$$\mathbf{P} = \varepsilon_0(\varepsilon - 1)\mathbf{E} \tag{3.54}$$

$$\mathbf{M} = \mu_0(\mu - 1)\mathbf{H} \tag{3.55}$$

Let us compare these formulas with our formulas obtained for a particular case when alternate electrical and magnetic fields are produced by an electromagnetic wave propagating in the medium where $\varepsilon>1$ and $\mu>1$.

Forces derived from the analysis of momentums and Newton's laws

In accordance with Eqs (3.40), (3.48) and (3.49), the force densities of various types are given by

$$\begin{aligned} &\mathbf{ff}_1 = -grad(\varepsilon)\varepsilon_0 \frac{E^2}{2}, \\ &\mathbf{ff}_2 - grad(\mu)\mu_0 \frac{H^2}{2}, \\ &\mathbf{ff}_3 = [\frac{dD}{dt} \times B], \\ &\mathbf{ff}_4 = [D \times \frac{dB}{dt}] \end{aligned} \tag{3.56}$$

From mathematical and physical points of view, Eq. (3.56) looks more attractive than Eq. (3.53). Two parameters - polarization **P** and magnetization **M** - are absent. As is seen, these parameters were not required in our analysis.

Although the types of forces in Eq. (3.53) and Eq. (3.56) are the same, their magnitudes are described by different formulas. There are the following differences.

Instead of $\mathbf{P} = \varepsilon_0(\varepsilon - 1)\mathbf{E}$ in Eq. (3.54) we have $\mathbf{D} = \varepsilon_0 \varepsilon \mathbf{E}$ for $\mathbf{ff}_1$ or $\mathbf{D} = \varepsilon_0 \varepsilon (1 - 1/\varepsilon\mu)\mathbf{E}$ for the second option.

Instead of $\mathbf{M} = \mu_0(\mu - 1)\mathbf{H}$ in Eq. (3.55) we have $\boldsymbol{B} = \mu_0 \mu \boldsymbol{H}$ for $\mathbf{fff}_1$ or $\mathbf{B} = \mu_0 \mu (1 - 1/\varepsilon\mu)\mathbf{H}$ for the second option.

Instead of $\mu_0 \boldsymbol{H}$ in Eq. (3.53) we have $\boldsymbol{B} = \mu_0 \mu \boldsymbol{H}$ or $\mathbf{B} = \mu_0 \mu (1 - 1/\varepsilon\mu)\mathbf{H}$ for the second option.

Instead of $\varepsilon_0 \boldsymbol{E}$ in Eq. (3.53) we have $\mathbf{D} = \varepsilon_0 \varepsilon \mathbf{E}$ or $\mathbf{D} = \varepsilon_0 \varepsilon (1 - 1/\varepsilon\mu)\mathbf{E}$ for the second option.

The forces $\mathbf{ff}_1$ and $\mathbf{ff}_2$ are expressed by perfectly different formulas than forces $\mathbf{f}_1$ and $\mathbf{f}_2$. As is shown, $\mathbf{ff}_1$ and $\mathbf{ff}_2$ can be derived from electrostatics and magneto statics. Unlike $\mathbf{f}_1$ and $\mathbf{f}_2$, forces $\mathbf{ff}_1$ and $\mathbf{ff}_2$ do not depend on the direction of the fields and are determined only by the inhomogeneous of the medium.

There are two corrections for force $\mathbf{f}_3$. First, force $\mathbf{f}_3$ must depend on inductance **B** rather than on the strength of magnetic field **H**. Second, the derivative of the displacement **D** should be used instead of the derivative of polarization **P**.

There are two corrections for force $\mathbf{f}_4$. First, force $\mathbf{f}_4$ must depend on displacement **D** rather than on the strength of electrical field **E**. Second, the derivative of the inductance **B** should be used instead of the derivative of magnetization **M**.

Thus, not one of the four expressions in Eq. (3.53) which are accordance with the Lorentz approach coincides with our expressions in Eq. (3.56).

The forces $\mathbf{f}_3$ and $\mathbf{f}_4$ are derived from general assumptions about the physical processes responsible for the occurrence of these forces. Different scientists have different assumptions, so there are several different formulas for the magnitude of these forces (Mansuipur, Zakharian and Wright 2013, 023826). In contrast, the magnitudes of our forces $\mathbf{ff}_1$, $\mathbf{ff}_2$ are obtained from generally accepted laws without any assumption about the natures of these forces.

Moreover, it is shown that similar types of forces, but of different physical natures, also exist for waves of other types. In the general case, forces arise when there is a change in space of two types of energies inherent in given wave. These energies exist in an arbitrary type of wave that moves through space at the speed of the wave. For an elastic wave, this is the kinetic energy arising from the oscillations of the medium in which the wave propagates, and the potential energy of compression/expansion of the medium. These energies are also equal in magnitude.

The nature of the forces arising from the propagation of an electromagnetic wave is different. These forces arise due to a change of electric and magnetic energies.

We consider a variant under the assumption that forces arise in free space for the same reasons as in an ordinary medium. In this case, there is no need to violate Newton's second law for free space. Indeed, can an electromagnetic wave propagate in free space? Yes, it can. Does the momentum of the wave change at some point? Yes, it does. Then why should there be no force at this point due to the change at this momentum? Another variant, in which there are no forces in free space, differs from the considered one only by the presence of a multiplier $(1\text{-}1/\varepsilon\mu)$ for forces $\mathbf{f}_3$ and $\mathbf{f}_4$ in Eq. (3.56).

Test of the approach based on the Lorentz forces

Let us calculate a magnitude of the density force based on the Lorentz force approach in the simplest particular case, when the electromagnetic wave incidents normally on the plane boundary of the dielectric, as is shown in

Fig. 3-1.

The two expressions for the force density based on the Lorentz forces approaches are known. We will consider the simplest case, where free charges and currents are absent and permeability μ=1.

The force density in accordance with Mansuripur, Zakharian and Wright (2013, 023826) is given by

$$f = (\boldsymbol{P}\square\nabla)\boldsymbol{E} + [\frac{d\boldsymbol{P}}{dt} \times \mu_0 \boldsymbol{H}] \quad (3.57)$$

Here $\boldsymbol{E}$ and $\boldsymbol{H}$ are respectively the electric and magnetic field strengths, and $\boldsymbol{P}$ is **the** polarization of the optical medium given by

$$\boldsymbol{P} = \varepsilon_0(\varepsilon - 1)\boldsymbol{E} \quad (3.58)$$

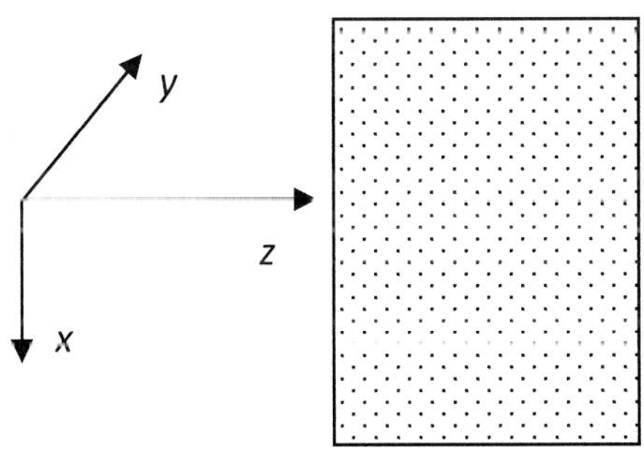

Fig. 3-1. Plane boundary between free space and optical medium.

The electrical and magnetic fields of the wave are directed along the x and y axes, respectively, and are given by

$$E_x = E, E_y = 0, E_z = 0; dE_x / dx = 0, \\ dE_y / dy = 0, dE_z / dz = 0 \quad (3.59)$$

$$P_x = \varepsilon_0(\varepsilon - 1)E_x, \ P_y = 0, \ P_z = 0;$$
$$dP_x / dx = 0, \ dP_y / dy = 0, \ dP_z / dz = 0 \tag{3.60}$$

$$H_x = 0, \ H_y = H, \ H_z = 0; \tag{3.61}$$

The first term in Eq. (3.57) describes the force that in accordance with Eqs (3.59), (3.60), and (3.61) is located in the *xy* plane. In this case the *z*-component of this force is equal to zero.

The second term in Eq. (3.57) describes the force, the *z*-component of which is different from zero, but where the phases d*P*/*dt* and *H* are shifted by $\pi/2$. As a result, the second term averaged over a period of oscillations is equal to zero (Torchigin and Torchigin 2014, 05780). Thus, Eq. (3.57) describes the force, the *z* component of which is equal to zero. This contradicts the results derived from thought experiments where no assumptions were made about the physical nature of optically induced forces. In addition, it has been known since Faraday's time that a force acting on a dielectric in an electric field is perpendicular to the surface of the dielectric, regardless of the direction of the electric field. This is also true for an alternating electric field produced by an electromagnetic wave.

Let us show that the pressure is nonzero in the same situation under the assumption that the momentum of light in the optical medium increases by a factor of η. Applied to the case under consideration in

Fig. 3-1, the MFD of the incident wave is divided into two parts. These are the MFD of the reflected wave and the MFD of the transmitted wave. In accordance with the Fresnel laws, the MFD of the reflection wave is given by

$$MFD_{REF} = \left(\frac{1-\eta}{1+\eta}\right)^2 MFD_{IN} \tag{3.62}$$

and the MDF_{TR} of the wave transmitted into the optical medium is given by

$$MFD_{TR} = \frac{4\eta}{(1+\eta)^2} x \tag{3.63}$$

Here *x* is the factor that characterizes an increase of the momentum in the optical medium. In accordance with thought experiments, $x=\eta$.

The pressure applied to the plane at $z=0$ is given by

$$P = MFD_{IN} - MFD_{REF} - MFD_{TR} = MFD_{IN}(1 + (\frac{1-\eta}{1+\eta})^2 - \frac{4\eta}{(1+\eta)^2}x) \tag{3.64}$$

At $x=\eta$, we have from Eq. (3.64)

$$P = MFD_{IN}\frac{2\eta^2 + 2 - 4\eta^2}{(1+\eta)^2} = -2\frac{\eta^2 - 1}{(1+\eta)^2} = -2MFD_{IN}\frac{\eta - 1}{\eta + 1} \tag{3.65}$$

Thus, the pressure produced by a wave at normal incidence from free space onto a flat boundary of the optical medium is negative, that is, a force arises on the boundary of the optical medium that is directed against the direction of the incident wave. This result was obtained by Thomson at the beginning of the 20th century (Thomson 1904).

On the contrary, in accordance with the Lorentz force approach the pressure applied to the plane is equal to zero. Taking into account this result and Eq. (3.64), we can calculate the unknown value x from the following equation

$$P = (1 + (\frac{1-\eta}{1+\eta})^2 - \frac{4\eta}{(1+\eta)^2}x) = \frac{2 + 2\eta^2 - 4\eta x}{(1+\eta)^2} = 0 \tag{3.66}$$

From Eq. (3.66) we have

$$x = \frac{\eta^2 + 1}{2\eta} = \frac{1}{2}(\eta + 1/\eta) \tag{3.67}$$

This value for the increase of the momentum in an optical medium was obtained by Mansuripur (2010, 1997-2005). Thus, Mansuripur, as well as ourselves, came to the conclusion that the momentum in an optical medium increases by a factor given by Eq. (3.67). This result was derived on the assumption that the pressure produced by a light wave at normal incidence on the plane surface of an optical medium is equal to zero. In fact, according to thought experiments the pressure is negative and is given by Eq. (3.65). In this case the momentum in an optical medium increases by η times.

Concluding remarks

Bearing in mind the fact based on many thought experiments that the momentum of a wave in a medium changes inversely with the speed u of its propagation, we have shown that the density of forces arising in a matter under the action of an electromagnetic wave is expressed by the same formulas as the time-averaged force density in electrostatics and magneto statics, provided that the magnetic and electric fields are variable and change sinusoidally.

Formulas for the density of forces in a homogeneous matter produced by an electromagnetic wave are also obtained. It is shown that there are two forces associated with a change in space and time in the energy densities of the electric and magnetic fields. The forces are equal in magnitude.

Similar forces must also exist in free space. Otherwise, such a case is an exception to the general rule and the magnitudes of forces in the general formulas must be multiplied by $(1\text{-}1/\varepsilon\mu)$.

The obtained formulas are compared with the known formulas obtained on the basis of the Lorentz force approach. As a result of the comparison, it was found that the number of various types of forces in both cases is the same (four). Moreover, the nature of these forces in both cases is the same. However, the formulas for all four types of forces are different. The formulas obtained by means of the analysis of a change in the momentum of the wave are derived from generally accepted laws without any assumptions about the nature of these forces. The formulas obtained by means of the Lorentz forces are derived under various assumptions. This is evidenced by the various versions of these formulas that currently exist. A list of the corrections that must be made to the formulas based on Lorentz forces in order to eliminate contradictions is given.

To make sure that the formulas based on the Lorentz forces give an incorrect result, the simplest particular case is considered, in which the pressure exerted by a light wave during normal incidence on a flat boundary between free space and an optical medium is calculated. The result obtained on the basis of the Lorentz forces is compared with the result obtained on the basis of a completely different approach, which does not use the formulas mentioned above. It is shown that the pressure on the boundary according to the formulas based on the Lorentz forces is zero, while the pressure based on the other approach is negative. The increase in the index of the momentum of light in an optical medium is calculated under the condition that the pressure on the boundary is zero. It turns out that this increase is exactly the one indicated by the authors of calculations based on the Lorentz forces. This indicates that zero pressure on the boundary leads to the incorrect conclusion that the increase in the momentum of light in the optical medium is different from n, where n is the reflective index of the optical medium.

From the above consideration, we can conclude that the result taken as a basis, which is valid for waves of an arbitrary type, according to which the wave momentum is inversely proportional to its propagation velocity, leads to reliable results regarding the forces arising in the matter under the action of an electromagnetic wave.

Optically induced forces responsible for mysterious ball lightning behavior

We have shown (Torchigin 2022) that it is the optically induced forces acting from the surrounding air on the light circulating in ball lightning that are responsible for all the anomalies associated with the intriguing and mysterious behavior of ball lightning in the Earth's atmosphere. Objects with circulating light are affected by yet another type of force, the magnitude of which is comparable to, and sometimes significantly exceeds, other known types of forces. This type of force, arising from the interaction of light with matter, was predicted by Maxwell and received experimental confirmation at the beginning of the 20th century. However, the magnitude of these forces acting on the light propagating in a straight line, even at the maximum achievable light energy densities, is so small that significant efforts are required to fix the existence of these forces.

In accordance with Eq. (3.35), the density of forces acting on light from the optical medium at $\mu=1$ and $\varepsilon=n^2$ is given by

$$\mathbf{f} = W_0 \frac{grad(n)}{n} \tag{3.68}$$

Note that the direction of the force is determined by the direction of the gradient of the reflective index and is independent of the direction of light propagation

As follows from the conditions necessary for stable circulating light, the energy density in it is more than 10 GPa (Torchigin 2021, 166635). The density of forces acting on light also increases as compared with a light propagating linearly accordingly. In ball lightning, the circulating light and the thin spherical shell of highly compressed air are an indivisible whole. Therefore, under the action of forces applied to circulating light, not only the circulating light, but also the entire ball lightning is subject to the action of the forces.

Integrating the left and right parts of Eq. (3.68) over the volume of the shell of circulating light, we obtain the following relationship between the force F acting on the circulating light and the amount of energy $\mathcal{E}$ of the circulating light stored in the shell

$$\mathbf{F} = \mathcal{E}\,\frac{grad(n)}{n} \tag{3.69}$$

Let us estimate the magnitude of optically induced forces for typical circumstances in which ball lightnings are observed. At normal pressure, the reflective index of air is given by

$$n = 1 + \Delta n_N \tag{3.70}$$

where Δn_N =2.78 10^{-4}.

Let the circulating light be in the situation in which ball lightnings are usually observed. This is a constant normal atmospheric pressure and a slightly different temperature. Since Δn_N is proportional to the air density, and the air density is inversely proportional to temperature T, the dependence of n on the temperature is given by

$$n = 1 + \Delta n_N \frac{T_0}{T} \tag{3.71}$$

where T_0=300^0 K at normal conditions.

Taking into account Eq. (3.69) and Eq. (3.71), we have

$$F = \mathcal{E}\,\frac{grad(n)}{n} = \mathcal{E}\,\Delta n_N grad(\frac{T_0}{T}) = -\mathcal{E}\,\Delta n_N \frac{grad(T)}{T} \tag{3.72}$$

From Eq. (3.72), we have it that the force applied to the circulating light is equal to 2.9 10^{-3} N when the temperature changes by one degree at a distance of 1 m.

Let us now estimate the mass of the circulating light. The mass consists mainly of the mass of the compressed air. On the assumption that its density is comparable with that of water, we have $m = \rho 4\pi R^2 h = 3.142$ 10^{-4} kg. In this case the weight of the circulating light is equal to 3.142 10^{-3} N. Thus, in the case under consideration, the optically induced force acting on the circulating light is comparable with its weight. In other words, the acceleration under the optically induced force is comparable with the acceleration of gravity *g*.

Thus, even with an extremely small inhomogeneity of the air, at which the temperature changes by one degree at a distance of one meter, the optically induced force is comparable with the force of gravity.

There are always weak inhomogeneities in temperature, associated with different thermal conductivities and the heat capacity of objects on the ground and periodic daily temperature changes. As a result, there are optically induced forces that act on circulating light. These forces are comparable in magnitude, and often exceed other known types of forces. In addition, it must be borne in mind that the circulating light, approaching some objects, heats them up due to its radiation. In turn, these objects heat the layers of air around them. As a result, the temperature of the air increases towards the object, and the gradient of the reflective index is directed in the opposite direction, that is, away from the object. Moving in the direction of the reflective index gradient, the circulating light cannot approach the object and bypasses it. It can be said that circulating light is the most sensitive device that reacts to the slightest inhomogeneities in the air temperature in the ambient space.

The above is true if the air pressure in the space under consideration is the same. This is true for short distances. It is known that air pressure decreases with height, despite the fact that the air temperature also decreases. Therefore, ball lightnings that arise at a height as a result of a linear discharge fall to the ground. However, they do not hit the surface of the earth, but stop at a distance of several meters from the surface. This confuses scientists, who cannot imagine how ball lightnings can know the distance to the earth's surface or what means they use to slow down and reduce their vertical speed to zero.

The explanation is simple. Ball lightning stops at the point where the gradient of the reflective index is equal to zero. It shows scientists that air density decreases as altitude increases above a few meters. At the same time, it shows that the density of air decreases as it approaches the surface of the earth from a distance of several meters to the very surface of the earth. Scientists need to explain only the second circumstance. The explanation is simple. Solar radiation heats the surface of the earth, and the surface of the earth heats the nearby layers of air due to the phenomenon of heat conduction. At the same time, the temperature of these layers increases, and the air density decreases. It is known that warm air is lighter than cold air, that is, the density of thermal air is less than that of cold air. This circumstance is used in aeronautics. Thus, due to the combination of these two circumstances, there is a maximum air density (maximum reflective index) at some distance from the surface of the earth. Ball lightning shows at what distance this maximum is located in each specific case.

A similar situation takes place in the case when ball lightning accompanies a flying airliner (Torchigin and Torchigin 2017, 196). Ball lightning is usually located on the edge of the wing. Scientists cannot explain why ball lightning is not blown away by the airstream there, which speed is greater than that of the strongest hurricane. Ejector seats are specially designed so that the pilot can leave the plane without being blown away by such strong winds. The explanation is simple. Again, ball lightning is located in the place where the air pressure is at its maximum. At the slightest deviation from this position, a reflective index gradient occurs, which leads to the appearance of an optically induced force directed towards the original position.

The action of optically induced forces also explains the fact that ball lightning is attracted to the flying airliner from all directions. A flying airliner disrupts the uniformity of the air around it. The air pressure is at its maximum in the place where the airliner is usually located. Around this place, the pressure gradually decreases. In this case, a volume of non-uniform air is formed around the flying airliner. The volume moves along with the airliner. Once in this volume, ball lightning moves in the direction of the air density gradient, that is, towards the airliner. This explains the fact that ball lightnings are observed by flying airliners and not airliners parked at airports.

Optically induced forces are also responsible for the amazing properties of ball lightnings that allow them to find holes and slots when penetrating into rooms (Torchigin 2019, 163126) and for the ability of ball lightning to change its spherical shape when penetrating into rooms through holes whose dimensions are smaller than the diameter of ball lightning (Torchigin and Torchigin 2016, 6155).

Thus, the optically induced forces, the existence of which has been known for more than a century and which until now have not been taken into account due to their extreme smallness, for circulating light increase to such an extent that their action becomes decisive.

CHAPTER FOUR

COMPRESSION PRODUCED BY AN ELECTROMAGNETIC WAVE

In every branch of knowledge, the progress is proportional to the number of facts on which to build, and therefore to the facility of obtaining data.
J. Maxwell

The circulating light causes compression of the air in the thin spherical shell in which it circulates due to a phenomenon that is closely related to optical electrostriction. This phenomenon is a special case of a more well-known phenomenon called electrostrictive pressure. As the name implies, this phenomenon is associated with a striction or compression of a body in an electrical field. Usually, a dielectric is subject to this phenomenon. This phenomenon was discovered by Helmholtz in 1878 for solid or liquid dielectrics and is different from another phenomenon connected with arising forces applied to the dielectric located in a static electric field.

Since the magnitude of this effect is proportional to the square of the electric field strength, this effect also takes place in an alternating electric field, in particular one produced by an electromagnetic wave and by light. The physical nature of this phenomenon is such that when light propagates in an optical medium, part of the light energy passes into the compression energy of the optical medium, which increases its reflective index. Such an increase contributes to a decrease in the energy of light. At equilibrium in a stationary state, there is a minimum of the total energy of the light and the compressed optical medium.

This immediately raises several questions. Is such a transition always possible? If such a transition is possible for light, then the balance between the energy of light and the optical medium should be considered, and not the balance between the energy of the electric field of light and the optical medium. There are currently no answers to these questions. Below is an examination of these issues, as a result of which completely new regularities are established, for which the existing ones are only a separate special case.

Electrostriction pressure in liquids and solids

Faraday began studying the forces acting on a dielectric in 1837. An example of such forces is the force acting on pieces of paper from the electric field produced by an electrified comb. Later, it turned out that the dielectric in an electric field was subject to the action of another type of force, the vector sum of which was zero, but which subjected the dielectric to compression. This phenomenon was called "the electrostriction pressure on a dielectric located in a static electric field" and was studied by Helmholtz in 1886 (Landau, Lifshits, Pitaevski 1976).

The generally accepted magnitude of the electrostriction pressure P_E produced by the electrostatic field of the strength E is given by

$$P_E = \varepsilon_0 \rho \frac{d\varepsilon}{d\rho}\frac{E^2}{2} \tag{4.1}$$

where ρ is the density of the dielectric and ε is its permittivity.

In order to use the same approach in the future for the analysis of compression in a field of electromagnetic waves, we first derive this formula, keeping in mind that the density of electric energy in a dielectric is determined follows

$$W_E = \varepsilon_0 \varepsilon \frac{E^2}{2} = \varepsilon W_{E0} \tag{4.2}$$

where W_{E0} is the energy density in free space where ε=1.

In this case, Eq. (4.1) can be presented as follows

$$P_E = \rho \frac{d\varepsilon}{d\rho} W_{E0} = \frac{\rho}{\varepsilon}\frac{d\varepsilon}{d\rho} W_E \tag{4.3}$$

Here P_E is the additional pressure produced by the electrical field. Note that the dimensions of pressure P_E [N/m^2=N m/m^3=J/m^3] and energy density [J/m^3] are the same. The dimensionless factor $\frac{\rho}{\varepsilon}\frac{d\varepsilon}{d\rho}$ determines the ratio between them.

Let us present the derivation of Eq. (4.3) to make sure that the approach we will use in what follows is the correct one.

Since the mass $m=\rho V$ of the body is unchanged, after taking logarithms and differentiating this expression, we have

$$\frac{dV}{V}=-\frac{d\rho}{\rho} \tag{4.4}$$

The relation between the energy of the electrical field of the strength E and the energy $\mathcal{E}$ stored in solid or liquid dielectric and its permittivity ε can be derived from the following considerations. The energy of a plane capacitor of C capacitance with charge q is given by

$$\mathcal{E}_E=\frac{q^2}{2C} \tag{4.5}$$

where

$$C=\varepsilon_0\varepsilon\frac{S}{h} \tag{4.6}$$

Here S is the area of plates and h is the distance between them. From Eq. (4.5) and Eq. (4.6) we have

$$\mathcal{E}_E=\frac{q^2 h}{2\varepsilon_0\varepsilon S} \tag{4.7}$$

If we increase permittivity ε, then in accordance with Eq. (4.7) the energy $\mathcal{E}$ stored in the capacitor and, therefore, in the dielectric decreases. Since others parameters in Eq. (4.7) are unchanged, we have the following invariant

$$\mathcal{E}_E\varepsilon=const \tag{4.8}$$

Again, taking logarithms and differentiating this expression, we have

$$\frac{d\mathcal{E}_E}{\mathcal{E}_E}+\frac{d\varepsilon}{\varepsilon}=0 \tag{4.9}$$

When a dielectric is placed in an electric field, then part of the energy of the electric field can go into another form, namely, into mechanical energy that occurs when the dielectric is compressed. Such a transition will continue until equilibrium is reached. It is known that equilibrium occurs at a minimum of total energy. In this case, the derivative of the sum of electrical and mechanical energies must be equal to zero, that is, the following condition must be satisfied

$$\frac{d\mathscr{E}_M}{dV} + \frac{d\mathscr{E}_E}{dV} = 0 \tag{4.10}$$

Here $\mathscr{E}_M$ is the energy of the dielectric as a result of its compression.

The change $d\mathscr{E}_M$ in the energy of a compressed body that is compressed by external pressure P with a change in the volume of the body by dV is determined as follows

$$\frac{d\mathscr{E}_M}{dV} = -P \tag{4.11}$$

In the case under consideration, the body is compressed by the ambient air pressure P_0 and the electrostriction pressure P_E produced by the electric field of the energy density W_E. Then, the first term in Eq. (4.10) is equal to the total pressure $-P=-(P_0+P_E)$. This term is negative because the energy always decreases with an increase of the volume. Taking into account Eq. (4.4) and Eq. (4.9), we can write Eq. (4.10) as follows

$$-(P_E + P_0) + \frac{\mathscr{E}_E}{\varepsilon}\frac{d\varepsilon}{d\rho}\frac{\rho}{V} = 0 \tag{4.12}$$

Since $\mathscr{E}_E / V$ is the density of the electrical energy, Eq. (4.12) can be rewritten as follows

$$P_E = \frac{\rho}{\varepsilon}\frac{d\varepsilon}{d\rho}W_E - P_0 \tag{4.13}$$

Comparing Eq. (4.3), with Eq. (4.13), we can conclude that they are the same if the liquid or solid is located in a vacuum at zero ambient air pressure

P_0. If the first term is smaller than the second one in Eq. (4.13), then there is no electrostriction pressure.

Thus, we can conclude that our approach based on an analysis of the steady-state, at which the minimum of the total energy takes place, leads to the correct result. At the same time, we found that the well-known Eq. (4.1) is valid under the assumption that the pressure of the air surrounding the liquids or solids is zero. If this pressure is different from zero, then the electrostriction pressure does not arise at low energy densities W_E of the electric field.

Electrostriction pressure in gases

Is Eq. (4.3) valid for gases? The answer is no. Eq. (4.3) is valid for objects in a vacuum, where the pressure of the external environment on the object is zero. An object in the form of a gas is under the pressure of the external environment, otherwise it would expand indefinitely. To obtain a steady state in a gas, it is necessary that the first term in Eq. (4.13) be equal to the second. This is only possible if the energy density of the electric field W_E is extremely high.

Actually, the dimensionless parameter $\frac{d\varepsilon}{d\rho}\frac{\rho}{\varepsilon}$ in Eq. (4.13) can be presented in the following form $\frac{(\Delta\varepsilon)/\varepsilon}{(\Delta\rho)/\rho}$. This is the ratio of the relative change of the permittivity $(\Delta\varepsilon)/\varepsilon$ to the relative change of the density $(\Delta\rho)/\rho$. This ratio is extremely small for gases at normal pressure. For example, if the air density at normal conditions increases by two times and $(\Delta\rho)/\rho=1$, then $(\Delta\varepsilon)/\varepsilon$ increases by $\Delta\varepsilon = 5.8\cdot 10^{-4}$ only.

Dependence of the reflective index n of the air density can be determined from the following Clausis-Mossotti relation

$$\frac{\varepsilon-1}{\varepsilon+2}=K\rho \qquad (4.14)$$

where K is a constant for each given substance.

Taking logarithms and differentiating both parts of Eq. (4.14), we get

$$d\varepsilon(\frac{1}{\varepsilon-1}-\frac{1}{\varepsilon+2})=\frac{d\rho}{\rho} \tag{4.15}$$

After transformations of Eq. (4.15), we have

$$\frac{3d\varepsilon}{(\varepsilon-1)(\varepsilon+2)}=\frac{d\rho}{\rho} \tag{4.16}$$

or

$$\frac{d\varepsilon}{d\rho}\frac{\rho}{\varepsilon}=\frac{(\varepsilon-1)(\varepsilon+2)}{3\varepsilon} \tag{4.17}$$

If the following condition

$$(\varepsilon-1)<<1 \tag{4.18}$$

is valid, then from Eq. (4.17) we obtain

$$\frac{d\varepsilon}{d\rho}\frac{\rho}{\varepsilon}=(\varepsilon-1) \tag{4.19}$$

It is known that for air under normal conditions $P=P_N=10^5$ N/m^2 and $\rho=\rho_N=1{,}2$ kg/m^3, the air permittivity is given by $\varepsilon=1+\Delta_N\varepsilon$, where $\Delta_N\varepsilon=5.4\ 10^{-4}$.

In this case, Eq. (4.19) can be presented as follows

$$(\varepsilon-1)=\Delta_N\varepsilon\frac{\rho}{\rho_N} \tag{4.20}$$

We can assume that the density of the air ρ is proportional to its pressure P_0 in the situation where Eq. (4.18) is valid. In this case Eq. (4.20) can be presented as follows

$$(\varepsilon-1)=\Delta_N\varepsilon\frac{P_0}{P_N} \tag{4.21}$$

Taking into account Eqs (4.19) and (4.21), we can present Eq. (4.13) as follows

$$P_E = \Delta_N \varepsilon \frac{P_0}{P_N} W_E - P_0 = P_0 (\frac{W_E}{P_N} \Delta_N \varepsilon - 1) \qquad (4.22)$$

As can be seen at the small energy densities of the electric field, the electrostrictive pressure is negative, which contradicts the accepted initial assumptions. This means that the electrostriction pressure is impossible at the small energy density W_E. In this case a small increase in gas density requires a greater energy than the electric energy can release when the permittivity ε increases due to this small increase of the gas density.

It should be noted that the threshold of the energy density of light W_{ET}, above which the electrostriction pressure in gas appears, is extremely great and is given by

$$W_{ET} = \frac{P_N}{\Delta_N \varepsilon} \qquad (4.23)$$

Since $P_N = 10^5$ Pa and $\Delta_N \varepsilon = 5.47\ 10^{-4}$, we have $W_{E0} = 1.8\ 10^8$ J/m^3. For comparison, the energy density of the electric energy in a high voltage capacitor of 100 mkF, 1 kV voltage and 1 cubic decimeter volume is equal to 0.5 10^5 J/m^3.

We can conclude from Eq. (4.22) that the electrostriction pressure P_E in gases arises if the energy density W_E of the electrical field is greater than a certain threshold.

Electromagnetic striction pressure in liquids and solids

Usually, when calculating the striction pressure produced by light in an optical medium, Eq. (4.1) is used to determine the striction pressure produced by a constant electric field. Since this pressure is proportional to the square of the electric field, such pressure can be produced by an alternating electric field, in particular, by an alternating electric field of a light wave. However, the energy of a light wave includes the energy of an alternating magnetic field, which should be taken into account. In this case we need to determine the balance between the energy of the light wave and the mechanical energy in the optical medium arising from the electromagnetic striction pressure (unlike electrostriction) produced by the light wave.

The total energy of the system in our case consists of the mechanical energy of the compressed optical medium under the action of the compression produced by the light wave and the energy of the electromagnetic field of the light wave. Knowing the parameters of the compressed optical medium at the steady state, we can determine the compression experienced by the optical medium located in the electromagnetic field of the light wave.

As with Eq. (4.8), we need to determine an invariant in this case. Consider a closed lossless system consisting of an optical medium placed in a field of an electromagnetic wave. Let there be a plane optical resonator filled with a homogeneous optical medium with permittivity ε and permeability μ, as is shown in

Fig. 4-1.

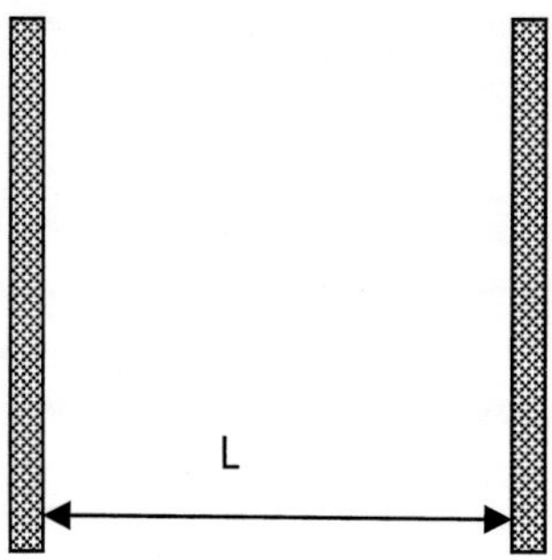

Fig. 4-1. Optical plane resonator filled by an optical medium.

The eigen frequencies of the resonator ω are determined from the condition that the phase shift at the bypassing of the resonator is given by

$$\frac{\sqrt{\varepsilon\mu}\omega}{c}L = 2\pi N \tag{4.24}$$

where L is the distance between reflectors, N is any integer, and $\sqrt{\varepsilon\mu}$ is a ratio of the speed of the wave in free space to that in the optical medium and can be considered as a generalized reflective index.

On the assumption that the generalized reflective index $\sqrt{\varepsilon\mu}$ is gradually changed in the resonator, the resonance frequency is also changed in accordance with Eq. (4.24). In this case we have

$$\sqrt{\varepsilon\mu}\omega = const \tag{4.25}$$

It is known that the resonance frequency ω is proportional to the energy $\mathscr{E}_{EH}$ stored in the resonator, and the following relation takes place

$$\frac{\omega}{\mathscr{E}_{EH}} = const \tag{4.26}$$

Bearing in mind that the volume in the case under consideration does not change, we have from Eq. (4.25) and Eq. (4.26) the following invariant

$$W_{EH}\sqrt{\varepsilon\mu} = const \tag{4.27}$$

where W_{EH} is the energy density of the electromagnetic field.

Taking into account that the total energy stored in solids and liquids is independent on the source of these energies, the final expression for the electromagnetic striction pressure P_{EH} produced in liquids or solids by the electromagnetic wave of W_{EH} energy density, unlike Eq. (4.13), is given by

$$P_{EH} = W_{EH0}\rho\frac{d\sqrt{\varepsilon\mu}}{d\rho} - P_0 = W_{EH}\frac{\rho}{\sqrt{\varepsilon\mu}}\frac{d\sqrt{\varepsilon\mu}}{d\rho} - P_0 \tag{4.28}$$

It should be pointed out that the term "optical electrostriction pressure", which is used to designate the striction pressure produced by light in an optical medium, does not correspond to the physical nature of this phenomenon. Indeed, the striction pressure will also be produced in the case when $\varepsilon=1$ and $\mu>1$. Therefore, after further consideration, we will use the new term "electromagnetic striction pressure".

Comparing Eq. (4.3) and Eq. (4.28), we can see that they are similar at $\mu=1$. But permittivity ε in Eq. (4.3) should be substituted by the reflective index since $n=\sqrt{\varepsilon}$. However, the energy density of the electrical field should be substituted by the energy density of the electromagnetic field. In this case, Eq. (4.28) can be presented as follows

$$P_{EH} = W_{EH} \frac{\rho}{n} \frac{dn}{d\rho} - P_0 \qquad (4.29)$$

Comparing Eq. (4.2) and Eq. (4.29), on assumption that $\varepsilon=n^2$, we can see that

$$\frac{\rho}{\varepsilon} \frac{d\varepsilon}{d\rho} = \frac{\rho}{n^2} \frac{dn^2}{d\rho} = \frac{\rho}{n^2} \frac{2ndn}{d\rho} = 2\frac{\rho}{n} \frac{dn}{d\rho} \qquad (4.30)$$

Thus, the multiplier in Eq. (4.2) is 2 times greater than that in Eq. (4.29). But the energy density W_{EH} is also greater by 2 times than the energy density W_E because the energy density of the magnetic wave is included in W_{EH} and this energy density is equal to the energy density of the electrical component of the electromagnetic wave.

Electromagnetic striction pressure in gases

Taking into account Eq. (4.22) for electrostriction pressure in gases, we can conclude that the electromagnetic pressure in gases can be presented as follows

$$P_{EH} = P_0 (\frac{W_{EH}}{P_N} \Delta_N \sqrt{\varepsilon\mu} - 1) \qquad (4.31)$$

where

$$\Delta_N \sqrt{\varepsilon\mu} = \sqrt{\varepsilon\mu} - 1 \qquad (4.32)$$

The threshold of the energy density of the electromagnetic field is given by

$$W_{EHT} = \frac{P_N}{\sqrt{\Delta_N \varepsilon\mu}} \qquad (4.33)$$

If $\mu=1$, then

$$\Delta_N \sqrt{\varepsilon\mu} \approx 0.5 \Delta_N \varepsilon \qquad (4.34)$$

Again, we need to note that the boundary density of the energy of light W_{EHT}, above which the electrostriction pressure in gas can arise, is extremely great. This great energy density can only be achieved in a field of circulating light.

Concluding remarks

We have considered the following 4 cases:

1. Striction pressure for solids and liquids in a static electrical field.
2. Striction pressure for gases in a static electrical field.
3. Striction pressure in solids and liquids in the field of an electromagnetic wave.
4. Striction pressure for gases in the field of an electromagnetic wave.

In the first section, we showed that the formula we obtained for the striction pressure in solids and liquids in a static electrical field completely coincides with the well-known and generally accepted Helmholtz formula known since the 19th century. But this formula is valid on condition that the liquids or solids are located in vacuum. At normal atmosphere pressure, a striction of the liquids and solids is absent at the small electrical energy density. Nevertheless, this indicates that the approach we adopted for calculating the striction pressure is correct and can be used for other cases.

In the second section, we used this approach to calculate the striction pressure in gases in a static electrical field. It is shown that the generally accepted formula is incorrect for gases. The striction pressure in gases arises at extremely great energy densities of the electrical field.

In the third section, we applied our approach to calculating the striction pressure in solids and liquids in the field of an electromagnetic wave. It is shown that the well-known formula correctly determines the striction pressure under the assumption that instead of the energy of the static electric field, one should understand the energy density of the electromagnetic wave, and the permittivity ε should be substituted by the generalized reflective index $\sqrt{\varepsilon\mu}$. In addition, ambient air pressure P_0 must be taken into account.

In the fourth section, we applied our approach to calculating the striction pressure in gases in the field of an electromagnetic wave. Again, the generally accepted formula is incorrect. The striction pressure in gases arises at extremely great energy densities of the electromagnetic wave.

The striction pressure is absent in all cases at a low energy density of the field that produces this pressure. Why is the threshold for the appearance of striction pressure for liquids and solids much smaller than for gases? It depends on how the mechanical and electromagnetic energies of an object change when it is compressed.

A product of the permittivity and permeability of liquids or solids are noticeably greater than unity. On the contrary, for gases this product differs from unity in the fourth decimal place at normal gas pressure. Therefore, for a gas, this product changes 10^4 times more slowly than for a liquid or solid. A change of the electromagnetic energy in a gas should be the same as in liquids or solids. From this requirement, it follows that a value of this energy in gases should be 4 orders of magnitude greater than in liquids or solids. This is true when the density of gases is much smaller than the density of liquids or solids. When a gas is compressed by a factor of 1000, its density increases by a factor of 1000 and becomes comparable to the density of liquids or solids. The same is true for the reflective index.

It is this situation that takes place in a spherical shell of highly compressed air, in which an intense white light circulates in all possible directions. The energy density of the circulating light turns out to be billions of times greater than the energy density of similar light propagating in a straight line. This is due to the fact that on each turn, the light continues to propagate in the same volume. Therefore, the volume of the circulating light is billions of times smaller than the volume of the light propagating in a straight line.

Let us consider in more detail the properties of the air subjected to electromagnetic striction pressure, since, contrary to the usual air pressure, which is familiar from everyday practice, the electromagnetic striction pressure only arises with extremely high light intensities, which are unattainable in conventional light beams propagating linearly.

A magnitude of the conventional gas pressure characterizes a tendency of gases to expand or a tendency of the gas molecules to push off from each other. Usually, this tendency is confined by the walls of the vessel in which the gases are confined. The gas pressure is equal to the force that is applied from the side of the gas to the unit area of the vessel's walls.

On the contrary, the magnitude of the electromagnetic striction pressure applied to the gas characterizes a tendency of the gas molecules to attract to each other. In this case, if the gas pressure is equal to P_1, and the electromagnetic striction pressure is equal to P_2, the resulting gas pressure on the wall of the vessel is equal to P_1-P_2. If $P_1<P_2$, then the resulting gas pressure is negative. This means that the pressure on the walls of the vessel is equal to zero and the volume of the gas in the vessel is decreased in such a way that the gas pressure in the decreased volume increases and becomes equal to P_2. Thus, the gas subjected to the electromagnetic striction pressure

can be confined without any vessel if the magnitude of the electromagnetic striction pressure surpasses the magnitude of the conventional gas pressure. This situation takes place in the shell of circulating light.

We use the term gas pressure for the generally accepted characteristic of gas to expand and to exert conventional pressure on the walls of the vessel, which prevents its expansion. We are going to use the term electromagnetic striction for gas to denote the tendency of a gas located in the field of a light wave to decrease its volume and increase its density. However, the word "striction" is not suitable to denote the compression of gas by external forces. In our opinion, the most appropriate term for the nature of the phenomenon is "contra-pressure", which shows that term is opposite to the generally accepted term "pressure", which characterizes the tendency of a gas to expand. We will intentionally use this term to show that pressure and contra-pressure exert the opposite effects on the gas. This term indicates the reason of the appearance of this contra-pressure, when the following situation takes place. If contra-pressure is greater than the conventional gas pressure, then no vessel to confine gas is required. Moreover, the contra-pressure will compress the gas until the gas pressure becomes equal to the contra-pressure.

The phenomenon of contra-pressure plays a decisive role in the existence of self-confined circulating light. The phenomenon is responsible for the compression of the air and increasing its reflective index. The area with the increasing reflective index forms an optical waveguide that provides the circulation of light and prevents its radiation in free space. We tried to present an additional clarification of this phenomenon, because for many readers the phenomenon seems unusual, under which the gas can be confined not only by the walls of the vessel or by gravity, just as the air in the Earth's atmosphere is confined, but also by the intense light propagating in this gas. In this case, the light confines the gas, and the gas, in turn, confines the light that propagates through it. This situation takes place in a natural ball lightning.

Thus, the study of the optical nature of ball lightning led to the need to study in more detail the phenomenon in which contra-pressure arises in a static electric field and in the field of an electromagnetic wave. In order to verify the correctness of the chosen approach, we first calculated the electrostriction pressure in a liquid or solid produced by a static electric field. It is shown that the obtained expression coincides with the generally accepted formula known since the 19th century, but only if the liquids or solids are in a vacuum, and it is not experienced by the initial pressure produced by atmospheric air. If a liquid or a solid is in the air under normal conditions, then the contra-pressure does not arise at the low electric energy density.

When considering the contra-pressure produced by an electromagnetic wave in liquids or solids, it is shown that not only the electric energy density of the wave, but also the magnetic energy density is responsible for the contra-pressure. As for gases, the contra-pressure arises only at the extremely great energy density of an electromagnetic wave. Such an energy density can only be achieved with the help of the circulating light, in which the energy density can be increased by a factor of billions compared to the energy density of light propagating in a straight line. This is connected with the fact that the volume of circulating light is reduced in proportion to the number of revolutions of the circulating light when it arises.

Centrifugal pressure produced by the circulating wave

Natural ball lightning has provided us with an ideal model for studying circulating light in an object where the boundaries of circulating light are absent. Light circulates in the sphere in all directions. Moreover, the light intensities in all directions are the same. Indeed, the scattering of light is proportional to its intensity. Thus, the bulk of the scattered light after scattering continues to circulate in the shell in other directions. Therefore, if the light intensity is greater in some directions than that in other directions, the scattering of the light in these directions is greater than that in other directions and, therefore, the intensity decreases to a greater degree until the intensity becomes equal to the intensities in the other directions.

The magnitude of the total momentum is given by

$$M = n\mathcal{E} / c \tag{4.35}$$

where $\mathcal{E}$ is the energy of the circulating light.

The momentum of the circulating light changes its direction during the circulation process. By Newton's second law, a change in the momentum is caused by the force F applied to the object in which the momentum changes. In the case under consideration, this force acts on the circulating light from the side of the shell of compressed air. By Newton's third law, there is another force F_S, which acts from the side of the circulating light to the shell. This force is equal to the force F in magnitude and is opposite to it in direction. Thus, the total force applied to the entire shell is given by

$$F_S = \frac{dM}{dt} = (n\mathcal{E} / c)\Omega \tag{4.36}$$

where

$$\Omega = 2\pi \frac{c/n}{2\pi R} = \frac{c/n}{R}. \qquad (4.37)$$

Here R is the radius of the sphere and Ω is the angular frequency of the rotation of the light around the sphere. The force F_S is directed towards the centre of the sphere. The opposite force is directed from the centre. This force is applied to the optical medium where the light is propagating. In our case, the optical medium is compressed air.

Taking into account Eq. (4.36) and Eq. (4.37), the total force F_S is given by

$$F_S = (n\mathcal{E}/c)\frac{c/n}{R} = \frac{\mathcal{E}}{R} \qquad (4.38)$$

The total energy of the circulating light can be presented as follows

$$\mathcal{E} = \mathrm{W}_L V \qquad (4.39)$$

where W_L is the energy density of the circulating light and V is the volume of the spherical shell given by

$$V = 4\pi R^2 h \qquad .(4.40)$$

where h is the thickness of the shell.

Taking into account Eq. (4.38), Eq. (4.39) and Eq. .(4.40), the total force F_S can be presented as follows

$$F_S = \frac{W_L 4\pi R^2 h}{R} = W_L 4\pi R h \qquad .(4.41)$$

The pressure on the shell is given by

$$P_C = F_S / S \qquad (4.42)$$

where S is the area of the lateral surface of the shell given by

$$S = 4\pi R^2 \qquad (4.43)$$

Taking into account Eq.(4.41), Eq. (4.42) and Eq. (4.43), the pressure produced by the circulating light on the surface of the shell is given by

$$P_C = W_L h / R \tag{4.44}$$

We call this pressure centrifugal pressure. This pressure is applied to the shell of compressed air and tends to increase the radius of the shell. At the same time, this pressure provides a spherical form of the shell. On the assumption that h/R is about 10^{-4}, we obtain that $P_C=10^{-4}W_L$. If W_L=1GPa, $P_C=10^5$ Pa or 1 bar. By the way, the excess pressure within a soap bubble is about 10 Pa only.

A spinning drum in a washing machine creates a similar pressure on the water in a damp cloth while removing moisture from the cloth. This pressure, like the pressure of compressed air in a soap bubble, tends to increase the radius of the sphere. The soap film prevents compressed air from expansion. There is no similar film that prevents the shell of the compressed air from expansion.

The force F_S is evenly distributed along the surface of the shell and tends to expand the sphere, that is, to increase the radius of the sphere. As a result, the force F_S makes positive work, and, therefore, the light energy stored in the shell tends to decrease.

We are interested in the pressure within the volume of the shell of thickness h that characterizes a degree of stretching between molecules.

Let us determine the total force applied to the left hemisphere shown in Fig. 4-2. The total force is directed perpendicular to the plane of the cross-section that is along the horizontal axis in Fig. 4-2.

Then, the total force is a sum of elementary forces directed along the x-axis. The elementary force is given by

$$\delta F_L = W_L \delta S \text{ Cos}\,\gamma \tag{4.45}$$

where δS is the area of the elementary surface and γ is the angle between the line perpendicular to the elementary surface and the x-axis.

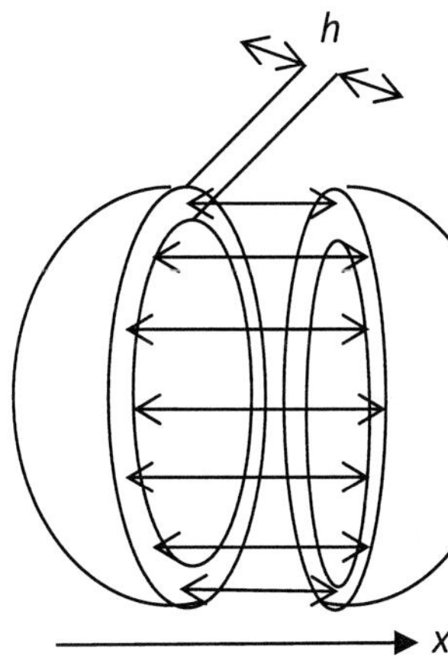

Fig. 4-2. Pressure P_L between two hemispheres of thickness h.

The magnitude $\delta S\ \mathrm{Cos}\,\gamma$ can be considered as a projection δSx of the elementary area δS at the plane perpendicular to the horizontal axis. In this case, the sum of all elementary projections is equal to the area of the circle of radius R. The force applied to the hemisphere is then given by

$$F_x = P_C \pi R^2 = W_L \frac{h}{R} \pi R^2 = W_L h \pi R \qquad (4.46)$$

The pressure P_{CS} applied to the ring in the cross-section of the $2\pi Rh$ area is given by

$$P_{CS} = F_x / (2\pi Rh) = W_L h \pi R / (2\pi Rh) = W_L / 2 \qquad (4.47)$$

This pressure tends to move the hemispheres apart. The pressure should be compensated by another pressure P_E of a different physical nature that tends to bring the hemispheres closer to each other. P_E should be equal to or greater than the pressure P_{CS}.

Eq. (4.47) can be obtained in another way using the following mathematical manipulations. The mechanical work produced by the pressure $P_{CS}\delta V$ is equal to a decrease of the energy $\mathcal{E}_L$ stored in the shell and the following relation is valid

$$P_{CS} = -\frac{d\mathscr{E}_L}{dV} = -\frac{d\mathscr{E}_L}{dR}\frac{dR}{dV} \tag{4.48}$$

Here V is the volume of the shell and is equal to $V = 4\pi R^2 h$. Then

$$dR/dV = (dV/dR)^{-1} = (8\pi Rh)^{-1}. \tag{4.49}$$

Since the number of wavelengths λ is not changed with a change of radius R of the circulating light, then λ/R=const. On the other hand, $\lambda=2\pi c/\omega$ where ω is the angular frequency of the light. In this case ωR=const. Since an isolated system is considered where the number of photons is unchanged and the energy of each photon is proportional to its frequency, we have the following relation

$$\mathscr{E}_L R = const \tag{4.50}$$

where $\mathscr{E}_L$ is the energy of the circulating light. Taking the logarithm and differentiating this relation, we get

$$\frac{d\mathscr{E}_L}{dR} = -\frac{\mathscr{E}_L}{R}. \tag{4.51}$$

Taking into account Eq. (4.48), Eq. (4.49) and Eq. (4.51), the centrifugal pressure within the shell is given by

$$P_{CS} = \frac{\mathscr{E}_L}{R}(8\pi Rh)^{-1} = \frac{1}{2}\frac{\mathscr{E}_L}{4\pi R^2 h} = \frac{1}{2}W_L \tag{4.52}$$

As is seen, Eq. (4.52) and Eq. (4.47) are identical. It should be emphasized that the pressure P_{CS} does not depend on either R or h and is determined only by the energy density W_L of the circulating light.

Separation of gases in a field of circulating light

As was shown for anomalous glowing objects (AGOs) arising due to erosive gas discharges in the Nauer experiments, the spectrums of AGO radiation correspond to gases, the reflective index of which are greater than that of the air (Barry 1980). Thus, once again, we can discover a new physical phenomenon by simply analyzing the available experimental data, bearing in mind that the light circulates in the AGOs. A necessary increase of the reflective index in the shell can be archived not only due to an increase of the air pressure in the shell but also due to drawing into the shell other gases with the greatest reflective indexes. There is a suspicion that such a method of increasing the reflective index in the shell requires less energy than increasing the reflective index by compressing air.

Again, the explanation of this phenomenon is a task with a solution known in advance. Indeed, the energy required for the separation of one mole of the gas mixture at pressure P_0 consisting of N components is given by

$$\mathcal{E} = RT\sum_{i=1}^{N} x_i \ln(\frac{P_0}{P_i}) \tag{4.53}$$

Here R=8.314 J mole^{-1} K^{-1}, T is the mixture temperature, x_i is the mole fraction of the i-th component, and P_i is the partial pressure of the i-th component where the mole fraction of the i-th component is given by

$$x_i = \frac{M_i}{M_1 + M_2 + \ldots + M_N} \tag{4.54}$$

Here M_i – is a magnitude of moles of the i-th component in the mixture. In a particular case of two components, the energy is given by

$$\mathcal{E} = RT[-x\ln(x) - (1-x)\ln(1-x)] \tag{4.55}$$

where x is the mole fraction of the first component and (1-x) is the mole fraction of the second component. The dependence of the energy on x is shown in
Fig. 4-3.

As is seen, the energy is maximal at x=0.5. In this case, numbers of the molecules of two components are identical. For example, the energy required to separate one mole of conventional air into nitrogen (79%) and oxygen (21%) is equal t o 1291 J/mole.

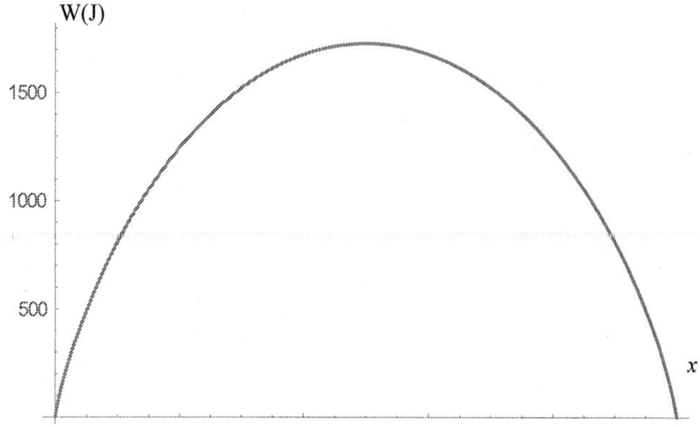

Fig. 4-3. The energy required for the separation of one mole of a two-component gas mixture where x is the mole fraction of one component.

As seen from

Fig. 4-3, the energy can be decreased by several orders of magnitude if a fraction of one component is extremely small. The volume of the shell is essentially smaller than the volume where the shell is located. It is explained by its very small thickness measured in micrometres. Then, a small percent of the component with great reflective index is required to increase the reflective index in the shell using the separation of the components and drawing into the shell the component with the greatest reflective index.

For example, if a fraction of ozone in the mixture is 10^{-4}, then in accordance with Eq. (4.55) the separation of the ozone requires 700 times less energy than if the fraction of ozone is 50%. Taken into account

Fig. 4-3, we can conclude that about 2 joules are required to separate the gas mixture where the fraction of ozone in the mixture is 10^{-4}.

The reflective index of ozone is $n = 1.00051$. The reflective index of air is $n = 1.00027$, which is 1.9 times smaller. For comparison, let us determine the energy that is required to compress the air by 1.9 times to get the same reflective index $n = 1.00051$.

The energy density W of the gas at pressure P is given by

$$W = P \tag{4.56}$$

Thus, the energy density of the air at normal conditions is equal to $W=10^5$ J/m^3. The energy density of the air at pressure $P=1.9\ P_0$ is equal to $1.9\ 10^5$ J/m^3. Thus, it requires $0.9\ 10^5$ J to compress 1 cubic meter of air 1.9

times. Then the energy that is required to compress one mole of air of volume 22.4 litres to obtain the same increase of the reflective index n=1.00051 is smaller by 0.0224 times and, therefore, is equal to 2016 J/mole.

Thus, on the assumption that the fraction of the ozone is extremely small, the energy required to increase the reflective index in the shell through drawing molecules of ozone into the shell is smaller by about three orders of magnitude than the energy required to obtain the same reflective index through compression of the air.

That is why an erosive gas discharge, in which a mixture containing hydrocarbons with a relatively high reflective index compared to air, favourably affects the AGO formation.

Furthermore, it should be borne in mind that the above considerations are related to the mixture at normal pressure and one mole of the mixture. With decreasing pressure, the amount of the mixture decreases, and the required energy required to separate the components is also reduced. At the same time, the energy required to compress air decreases accordingly. Therefore, with a decrease in pressure the ratio between the densities of light energy required to compress the air by 1.9 times and to separate molecules of ozone remain unchanged and are equal about three orders of magnitude.

Experiments of Nauer and Barry (Barry 1980) were more successful at low chamber pressures. We explain this effect as follows. The separation of the ozone molecule occurs from areas adjacent to the area where intense light exists. For complete separation of the mixture, molecules from more distant areas must profound into these areas. This takes time. The discharge time and hence the lifetime of intense light is only a few milliseconds. With a decrease in the pressure of the mixture, the mean free path increases, and a larger number of ozone molecules will penetrate the area where there is intense light during the discharge time. In other words, with a decrease in mixture pressure, the diffusion rate increases, and during the short time of the discharge, a larger number of ozone molecules enter the area of intense light and increase the reflective index in the area.

CHAPTER FIVE

STEADY-STATES AND THE STABILITY OF THE BALL LIGHT

There are more things in heaven and earth, Horatio,
Than are dreamt of in your philosophy
Hamlet - William Shakespeare

Introduction

We made several attempts to justify ball light stability. Judging by the reaction of the readers, these attempts were not convincing. Perhaps this is due to the fact that stability justification requires knowledge not so much of optics as mechanics.

The first thought that comes to a person when he hears about a circulating light, is what does it circulate around? When he hears that the light is circulating on his own without any additional restrictions, he becomes perplexed. How can this be? Light travels in a straight line. Why would it spin? Suppose that it spins in an inhomogeneous medium, and you are asked where can you get such an inhomogeneous medium so that its curvature is in the form of a circle. To the answer that the circulating light itself creates such an inhomogeneous medium in the air, perplexity increases even more.

To completely refute the idea of circulating light, such an irrefutable objection is given. Let the light circulate in the inhomogeneous air medium that it created for itself. But the air is not hard glass, the shape of which is independent of the circulating air. Under the influence of circulating light, this air structure will expand and cannot be stable. These objections are so convincing and obvious that for two centuries, no one has even voiced a combination of circulating light, not to mention the study of this phenomenon – ball light. The objections are fair and difficult to refute.

On the other hand, we were convinced that circulating light exists in nature, and there must be an explanation for this. As follows from the thoughts of famous scientists, if the phenomenon of nature contradicts existing knowledge, then the phenomena of nature should be given priority. Perhaps modern knowledge is insufficient to explain this phenomenon. Perhaps the knowledge is sufficient, but it cannot be applied properly. We found that the second assumption is correct. It turned out that modern knowledge is more than enough. Moreover, it turned out that knowledge available in the 18th century, when such concepts as a waves, energy, and reflective indexes became known, was quite sufficient.

Studying the anomalies and mysteries of the behaviour of ball lightning, we concluded that this was a completely new object that penetrated through window panels and could not consist of material particles, since they could not penetrate through glass.

Only light can penetrate glass. Ball lightning emits light, but there are no material particles in it. The question arises - how can light be kept in a limited volume in the form of a ball without material particles? The answer is, it is impossible in a vacuum, but it is possible in conventional air. A ball light model has appeared.

Further, there was no problem making sure that all the anomalies in the behaviour of ball lightning are explained by optically-induced forces, which act at the ball light.

The purpose of this chapter is to prove the stability of the circulating light and determine its physical parameters, for which such stability takes place. As already noted, we believe that the existence of circulating light in nature does not follow from theoretical considerations, but from the indisputable fact that the behaviour of circulating light in the earth's atmosphere, based on the laws of optics known since the 18th century, completely coincides with the mysterious and intriguing behaviour of natural ball lightning. Therefore, the development of the theory of circulating light is a task with a known answer in advance.

The analysis of the stability of an isolated conservative system is a task related more to mechanics than to optics, although circulating light plays the main role in this system. Since the circulating light, judging by the name, is basically an optical phenomenon, it is difficult for optical specialists to judge issues related to mechanics.

To facilitate understanding of the problems associated with circulating light, we first will consider the stability of the object, which is in many ways similar to circulating light, but whose stability is undeniable. This is a soap bubble, the stability of which, like the stability of circulating light, is determined by the balance of two types of energy. For a soap bubble, this is the energy of the compressed air and the stretched soap film. For a light bubble, this is the energy of the compressed air and the energy of the circulating light.

After the stability of the soap bubble is proved, one more object, in the form of a child's balloon, is considered to show that the same approach works in this case also. Knowing the concepts and methods used to analyze stability, we will apply the same approach, the same concepts, and the same notations to analyze the stability of the circulating light.

Fortunately, the mechanical system for a soap bubble and circulating light turned out to be very similar and simple. Both systems contain only one degree of freedom and only two types of energies. Moreover, one type of energy related to compressed air is the same in both systems. In fact, the analysis of the stability of the circulating light coincides with the analysis of the stability of a soap bubble. Earlier, when analyzing the circulating light, we made backlinks to the corresponding formulas for the soap bubble. Later, we decided that the forward links were also useful, as they show that the analysis of the stability of a ball light does not differ in complexity from the simplest analysis of the stability of a soap bubble. In our opinion, this should provide an additional argument for the reliability of the approach to the analysis of the fulfillment of the necessary conditions for the stability of the circulating light.

The stability analysis allows us to refute many seemingly irrefutable arguments that the necessary stability conditions are not met. But meeting the necessary conditions does not mean that sufficient conditions are met. The stability of the circulating light, in our opinion, follows from the fact that its behaviour in the earth's atmosphere coincides with the behaviour of another luminous object, which people call ball lightning. We regard this object as experimental confirmation of the existence of the circulating light and hence its stability.

Two approaches to an analysis of stability are possible. The first approach, let's call it mathematical, is connected with the preparation of mathematical equations that determine the behaviour of the system and the formulation of the necessary conditions under which the system will be stable. After that, it is shown through mathematical manipulations that these conditions are satisfied. In this case, the interaction of various physical phenomena is hidden because the laws of mathematics perform this function.

The second approach, let us call it physical, involves introducing into consideration the corresponding known physical phenomena and laws with the subsequent analysis of their interaction.

Each approach has its advantages and disadvantages, and its adherents and its opponents. In our opinion, the first approach prevails in modern theoretical physics.

We use both approaches. We believe that the second approach is necessary for explaining the physical nature of the circulating light since it explains the relationship between the new phenomenon under consideration and the known phenomena described in school textbooks.

In interpreting the results obtained through the first approach, it turned out that some of the phenomena used in the second approach require clarification, and some should be generalized. Thus, the analysis of the stability of circulating light is accompanied by the acquisition of new knowledge about phenomena that have independent significance.

In addition to the proof that ball light is stable, many other surprising and unexpected phenomena described in the previous chapter are applied. An attempt is made to explain these phenomena, taking into account the structure and properties of the circulating light.

Bearing this in mind, it is quite simple to explain all the physical parameters of the ball light, as well as to consider the features of its appearance and disappearance. Furthermore, the anomaly associated with the very existence of circulating light is explained. True, it is difficult to call it an anomaly because it is in agreement with well-known laws of optics and physics.

Steady states and stability of a soap bubble

Introduction

The notion of stability is much broader than the stability of circulating light. For systems with energy, there is a generally accepted approach to determining the fulfilment of the necessary conditions for the existence of stability. It is assumed that the system is characterized by some parameters that determine its state. If these parameters do not change over time, then such a state is called a steady-state. The steady-state is considered stable if the system remains in a steady state with small change to these parameters. This condition is necessary, but not sufficient. However, in many cases, the necessary conditions are not met, so further stability studies are not required.

Concerning circulating light, a stable steady-state is one in which its parameters, such as the radius and thickness of the shell, the air pressure in the shell, and the energy density of the circulating light, are invariable over time. Strictly speaking, there is no steady-state of the circulating light, since the energy of the circulating light gradually decreases due to the inevitable radiation losses. In theory, circulating light cannot exist indefinitely.

However, it can exist long enough to study its behaviour with small changes to its parameters.

Since the stability of circulating light has never been studied by anyone before, it makes sense as an introduction to show how the stability of a system is studied using simpler examples, so that you can be sure that the analysis of the stability of circulating light is performed in the same way as the analysis of the stability of other known systems.

The most similar to circulating light in shape and components is an ordinary soap bubble, an idea of which every reader has and the stability of which is beyond doubt (although the stability of a soap bubble is also conditional since it ultimately becomes unstable due to changes to the parameters over time of the soap film associated with water evaporation, etc.).

As is known, a steady-state of a closed conservative system is stable if its total energy is minimal at a change of any parameter from the steady-state value. We first consider how such requirements are satisfied for a soap bubble. Then, in the same way, we analyze the fulfilment of these requirements for ball light.

There are two types of energy in the soap bubble. These are the energy of the compressed air $\mathcal{E}_A$ and the energy of the stretched shell $\mathcal{E}_S$. The total energy of the bubble is given by

$$\mathcal{E}_S + \mathcal{E}_A \tag{5.1}$$

As is known, a minimum of any function with a change of any parameter x takes place at the point where its derivative over x is equal to zero. In this case

$$\frac{d\mathcal{E}_{TOTAL}}{dx} = \frac{d\mathcal{E}_A}{dx} + \frac{d\mathcal{E}_S}{dx} = 0 \tag{5.2}$$

From Eq. (5.2) we have

$$\frac{d\mathcal{E}_A}{d\mathcal{E}_S} = \frac{d\mathcal{E}_S}{d\mathcal{E}_A} = -1 \tag{5.3}$$

This means that an increase in the energy of one type should cause the same decrease in the energy of the other type at a steady state. We can conclude from Eq. (5.3) the following first necessary condition for the existence of a steady-state. At a slow change of any parameter x of the bubble $\mathscr{E}_S(x)$, the functions and $\mathscr{E}_A(x)$ should not be simultaneously either increasing or decreasing at the point of a steady-state. The fulfilment of this requirement for bubbles of light raises great doubts among specialists. The fulfilment of this requirement for the soap bubble should not be in doubt, as everybody knows that such bubbles exist. First, we consider how the stability of the soap bubble is proven. Then, similarly, we will analyze the stability of ball lights.

The simplest analysis of the stability of a soap bubble

The energy of the shell of the soap bubble is given by

$$\mathscr{E}_S = 2\eta S \tag{5.4}$$

where η [J/m^2] is the index of surface tension and S is the surface area of the bubble given by

$$S = 4\pi R^2 \tag{5.5}$$

where R is the radius of the bubble.

Taking into account Eq. (5.5) and Eq. (5.4), we have

$$\frac{d\mathscr{E}_S}{dV} = \frac{d\mathscr{E}_S}{dS}\frac{dS}{dV} = \frac{d\mathscr{E}_S}{dS}\frac{dS}{dR}\frac{dR}{dV} = 2\eta\frac{8\pi R}{4\pi R^2} = \frac{4\eta}{R} \tag{5.6}$$

where volume V of the bubble is given by

$$V = 4\pi R^3/3 \tag{5.7}$$

Let us now consider a change of the second type of energy in the form of the energy of the compressed air within the soap film. As is known, a change of the energy $\Delta\mathscr{E}$ of any compressed air by the volume ΔV is given by

$$\Delta\mathscr{E}_A = -P_E\Delta V \tag{5.8}$$

where P_E is the excess pressure exerted by the body. Taking into account Eq. (5.8), we have

$$\frac{d\mathcal{E}_A}{dV} = -P_E \tag{5.9}$$

Taking into account Eq. (5.6) and Eq. (5.9), we have from Eq. (5.2) the following condition of steady-state

$$4\eta / R_0 = P_E \tag{5.10}$$

This is a dependence of the radius R_0 of the soap bubble on the excess pressure P_E within the shell at a steady state. This is the known formula that determines the excess pressure within a soap bubble. We can see from Eq. (5.10) the interesting properties of soap bubbles. The greater the radius of a soap bubble, the smaller the air pressure within its shell.

Let us show that the steady-state in accordance with Eq. (5.10) is stable. For this purpose, we need to show that the derivative of the total energy over the volume (or over the radius, because the radius is an increasing function of the volume) is an increasing function of volume at $R=R_0$. We will designate parameters at the steady-state by index 0.

Taking into account Eq. (5.2), Eq. (5.6) and Eq. (5.9), the derivative of the total energy over the volume can be presented as follows

$$\frac{d\mathcal{E}_{TOTAL}}{dV} = \frac{4\eta}{R} - P_E(V) \tag{5.11}$$

In accordance with the Clapeyron relation, we have

$$P_E V = P_{E0} V_0 \tag{5.12}$$

Let us introduce the following dimensionless parameter that is proportional to V and is equal to the unit at a steady-state

$$\beta = \frac{V}{V_0} \tag{5.13}$$

where V_0 is the volume at a steady state.

We are interested in the parameters at a steady state when β=1. From Eq. (5.13), we can derive the following relations

$$\beta^{1/3} = \frac{R}{R_0} \quad and \quad \beta^{2/3} = \frac{S}{S_0} \tag{5.14}$$

Taking into account Eq. (5.6), Eq. (5.9), Eq. (5.13), and Eq. (5.14), we have

$$\begin{aligned} &\frac{d\mathscr{E}_{TOTAL}}{dV} = \frac{4\eta}{R(V)} - P_{E0}\frac{V_0}{V} = \frac{4\eta}{R_0}\beta^{-1/3} - P_{E0}\beta^{-1} = \\ &\frac{4\eta}{R_0}[\beta^{-1/3} - (\frac{4\eta}{R_0})^{-1} P_{E0}\beta^{-1}] = \frac{4\eta}{R_0}[\beta^{-1/3} - \beta^{-1}] \end{aligned} \tag{5.15}$$

Taking into account Eq. (5.10) and Eq. (5.15), the second derivative of the total energy over the volume is given by

$$\begin{aligned} &\frac{d^2\mathscr{E}_{TOTAL}}{dV^2} = \frac{d\beta}{dV}\frac{d}{d\beta}\{\frac{4\eta}{R_0}[\beta^{-1/3} \quad \beta^{-1}]\} = \frac{4\eta}{R_0}\frac{1}{V_0}\frac{d}{d\beta}[\beta^{-1/3} - \beta^{-1}] = \\ &\frac{4\eta}{R_0}\frac{1}{V_0}[-\frac{1}{3}\beta^{-4/3} + \beta^{-2}] \end{aligned} \tag{5.16}$$

As can be seen, the second derivative is positive at the steady-state, where β–1, and, therefore, the total energy has a minimum at β=1. This means that the steady-state is stable.

Another approach to analysis of stability of a soap bubble

When analyzing the stability of circulating light, the main parameter that characterizes the state of compressed air is the dimensionless parameter compression σ, which determines the degree of the air compression as compared with normal conditions. The compression is given by

$$\sigma = \frac{\rho}{\rho_0} = \frac{V_0}{V} \tag{5.17}$$

Here ρ and ρ_0 are densities of the compressed air and the air at normal conditions, respectively. V and V_0 are volumes of the compressed air and the same air at normal conditions. We are considering the compressed air where $\sigma>1$.

If the air pressure is smaller than 100 atmospheres, the following relation is valid

$$\sigma=\frac{P}{P_N} \tag{5.18}$$

where P and P_N are the pressure of the compressed air and the pressure at normal conditions.

Let us now consider a typical case of an analysis of the stability of a steady-state where a notion of compression is used. Parameters at a steady state will be designated by index 0. The compression, pressure, and radius at the steady-state are equal to σ_0, P_0, and R_0, respectively. The first derivative of the energy of the compressed air over the volume in accordance with Eq. (5.9) and Eq. (5.18) can be presented as follows

$$\frac{d\mathcal{E}_A}{dV}=-\sigma P_0 \tag{5.19}$$

From Eq. (5.13) and Eq. (5.17), we have

$$\sigma=\sigma_0/\beta \tag{5.20}$$

The first derivative of the energy of the shell over the volume in accordance with Eq. (5.6), Eq. (5.14), and Eq. (5.20) can be presented as follows

$$\frac{d\mathcal{E}_S}{dV}=\frac{4\eta}{R_0\beta^{1/3}}=\frac{4\eta}{R_0}\left(\frac{\sigma}{\sigma_0}\right)^{1/3} \tag{5.21}$$

In the steady state at $\sigma=\sigma_0$, parameter R_0 has a definite value, which is determined from the condition for the minimum total energy, that is, from the condition

$$\frac{d\mathcal{E}_{TOTAL}}{dV} = 0 \tag{5.22}$$

Thus, a dependence of the first derivative of the total energy over the compression can be presented instead of Eq. (5.11) as follows

$$\frac{d\mathcal{E}_{TOTAL}}{dV} = \frac{4\eta}{R_0}(\frac{\sigma}{\sigma_0})^{1/3} - \sigma P_0 \tag{5.23}$$

Taking into account (5.22), we have from Eq. (5.23) at $\sigma = \sigma_0$

$$\frac{4\eta}{R_0} = \sigma_0 P_0 \tag{5.24}$$

Taking into account Eq. (5.20), Eq. (5.21), and Eq. (5.24), we can write Eq. (5.23) as follows

$$\frac{d\mathcal{E}_{TOTAL}}{dV} = P_0\sigma_0[(\frac{\sigma}{\sigma_0})^{1/3} \quad \frac{\sigma}{\sigma_0}] \tag{5.25}$$

Taking into account Eq. (5.20), we have from Eq. (5.25)

$$\frac{d\mathcal{E}_{TOTAL}}{dV} = P_0\sigma_0(\frac{\sigma_0 / \beta}{\sigma_0})^{1/3} - \frac{\sigma_0 / \beta}{\sigma_0}) - P_0\sigma_0(\beta^{-1/3} - \beta^{-1}) \tag{5.26}$$

As can be easily seen, the second derivative of the total energy over the volume V or over β is positive, and therefore the first derivative is an increasing function of the compression. If we differentiate Eq. (5.26) over β, we obtain Eq. (5.16). This means that the steady-state is stable.

When studying the stability of circulating light, there is also a balance in the steady state of the volume derivatives of the energies of two types. This is the energy of circulating light, which creates the contra-pressure that prevents the expansion of compressed air in a spherical shell. This is the energy of compressed air, which, like any compressed gas, tends to expand. In the presented case, instead of the energy of circulating light, the energy of the stretched film and the same energy of the compressed gas are considered. Both energies have an analytical representation, so their stability can be determined analytically.

For circulating light, both types of energy do not have an analytical representation, since the densities of these types of energy are so high that the generally accepted formulas are inapplicable. In this case, the dependences of these energies are obtained not on the volume, but on the degree of air compression. Since the product of compression and volume is a constant value, an increase in volume is associated with a decrease in compression and vice versa. In this case, it is necessary to investigate the dependence of the derivative of the total energy $\frac{d\mathscr{E}_{TOTAL}}{d\sigma}$ over compression σ. Then, the steady-state is stable if the dependence of the total energy on compression is an increasing function of σ at the steady state. Let us show that this is indeed the case.

Dependence $\frac{d\mathscr{E}_{TOTAL}}{d\sigma}$ can be obtained as follows. From Eq. (5.17) we have

$$\frac{d\sigma}{dV} = -\frac{\sigma}{V} \tag{5.27}$$

Then

$$\frac{d\mathscr{E}_{TOTAL}}{dV} = \frac{d\mathscr{E}_{TOTAL}}{d\sigma}\frac{d\sigma}{dV} = -\frac{d\mathscr{E}_{TOTAL}}{d\sigma}\frac{\sigma}{V} = -\frac{dW_{TOTAL}}{d\sigma}\sigma \tag{5.28}$$

Taking into account Eq. (5.25), we have

$$\frac{d\mathscr{E}_{TOTAL}}{d\sigma} = -P_0\frac{\sigma_0}{\sigma}[(\frac{\sigma}{\sigma_0})^{1/3} - \frac{\sigma}{\sigma_0}] = P_0[1-(\frac{\sigma}{\sigma_0})^{-2/3}] = P_0[1-(\frac{\sigma_0}{\sigma})^{2/3}] \tag{5.29}$$

Since $(\frac{\sigma_0}{\sigma})^{2/3}$ is a decreasing function of σ, $[1-(\frac{\sigma_0}{\sigma})^{2/3}]$ is an increasing function of σ. Fig. 5-1 shows this function. As can be seen, $\frac{d\mathscr{E}_{TOTAL}}{d\sigma}$ is an increasing function of σ/σ_0.

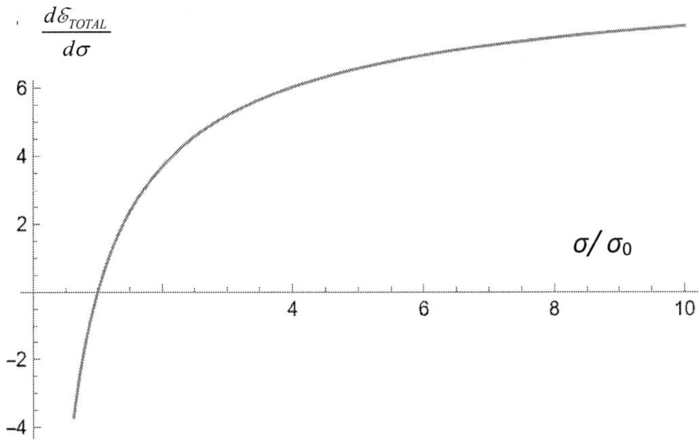

Fig. 5-1. Dependence of $\frac{d\mathcal{E}_{TOTAL}}{d\sigma}$ on the compression σ in a form σ/σ_0 where σ_0 is the compression at the steady-state.

For comparison, dependence $\frac{d\mathcal{E}_{TOTAL}}{dV}$ on parameter β is presented in Fig. 5-2.

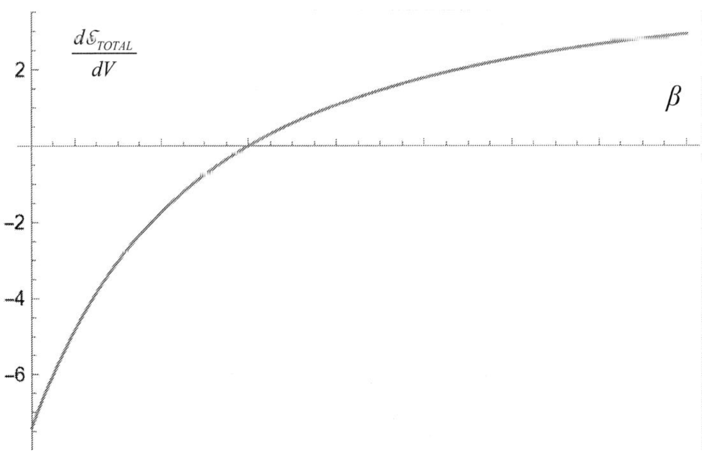

Fig. 5-2. Dependence of $\frac{d\mathcal{E}_{TOTAL}}{dV}$ on the volume in a form of parameter β where β=1 at steady state.

As is seen, in both case dependences, both $\frac{d\mathscr{E}_{TOTAL}}{d\sigma}$ and $\frac{d\mathscr{E}_{TOTAL}}{dV}$ are increasing functions.

Steady states and stability of an inflated child's ball

Let us now consider, following the given template, the stability of an ordinary children's ball. Let the air pressure at a steady-state be equal to P_0. The energy density of the stretched rubber shell, unlike the energy density of the soap bubble shell in accordance with Eq. (5.4), is given by

$$\mathscr{E}_S = K(S - S_N)^2 = K(4\pi R^2 - 4\pi R_N^2)^2 \tag{5.30}$$

where K is multiplier dependent on the thickness and properties of rubber, and $S_{N,}$ and R_N are the initial surface and initial radius of the shell, respectively. As can be seen, the energy of the stretched shell is proportional to the square of a change of the area of the shell.

In this case, instead of Eq. (5.4), we have

$$\frac{d\mathscr{E}_S}{dV} = \frac{d\mathscr{E}_S}{dR}\frac{dR}{dV} = \frac{2K}{4\pi R^2}(R^2 - R_N^2)8\pi R = \frac{4K}{R}(R^2 - R_N^2) \tag{5.31}$$

Let us assume that the initial conditions of our ball are the following. The shell has a spherical form of radius $R_{N,}$ and the air pressure within the ball is equal to the normal pressure P_N and therefore the air compression is σ_N=1. With an increase in compression $\sigma_0>\sigma_N$, the air pressure $P_0>P_N$. As a result, the radius of the ball at a steady state is $R_0>R_N$. Thus, the ball takes a new steady state.

As before, the dependence of the energy of the compressed air is given by

$$\frac{d\mathscr{E}_A}{dV} = -P_0 / \beta \tag{5.32}$$

where compression P_0 is the pressure at a steady state. Parameter β=1 at the steady state is given by

$$\beta = \frac{P_0}{P} = \frac{V}{V_0} \tag{5.33}$$

The derivative of the total energy over the volume of the ball of radius R_0 at a steady stare can be presented as follows

$$\frac{d\mathcal{E}_{TOTAL}}{dV}=\frac{d\mathcal{E}_S}{dV}+\frac{d\mathcal{E}_A}{dV}=\frac{4K}{R_0}(R_0^2-R_N^2)-P_0=0 \quad (5.34)$$

From Eq. (5.34), we have the following relations between parameters at the steady state

$$\frac{4K}{R_0}(R_0^2-R_N^2)=P_0 \quad (5.35)$$

If radius R is different from R_0, Eq. (5.34) can be presented as follows

$$\frac{d\mathcal{E}_{TOTAL}}{dV}=\frac{4K}{R_0\beta^{1/3}}(R_0^2\beta^{2/3}-R_N^2)-\beta^{-1}\frac{4K}{R_0}(R_0^2-R_N^2)=$$
$$\frac{4K}{R_0}[(R_0^2\beta^{1/3}-R_N^2\beta^{-1/3})-\beta^{-1}(R_0^2-R_N^2)] \quad (5.36)$$

To keep the steady state stable, the second derivative must be positive at β=1. Differentiating Eq. (5.36) with respect to β, we get at β=1.

$$\frac{d^2\mathcal{E}_{TOTAL}}{dV^2}=\frac{d}{dV}(\frac{d\mathcal{E}_{TOTAL}}{dV})=\frac{d}{d\beta}(\frac{d\mathcal{E}_{TOTAL}}{dV})\frac{d\beta}{dV}=\frac{d}{d\beta}(\frac{d\mathcal{E}_{TOTAL}}{dV})\frac{1}{V_0}=$$
$$\frac{4K}{R_0}[(R_0^2/3+R_N^2/3)+(R_0^2-R_N^2)]=\frac{4}{3}R_0^2-\frac{2}{3}R_N^2$$
$$(5.37)$$

Since $R_0>R_N$, the second derivative over the volume in Eq. (5.37) is positive. This means that $\mathcal{E}_{TOTAL}$ at the steady state is minimal and, therefore, the steady state is stable.

Concluding remarks

When analyzing stability, it is preferable to use the analysis of stability when changing the dimensionless parameter σ, which has a clear physical meaning and determines in accordance with Eq. (5.17) the degree of compression of air relative to air at normal pressure. However, the stability of a steady-state can be convincingly explained using the parameter that determines the volume, since when considering stability, it is usually objected that the circulating light is unstable and its volume increases indefinitely. Therefore, stability was demonstrated both with a change in volume and with a change in compression. In accordance with Eq. (5.27), $\frac{d\mathcal{E}_{TOTAL}}{dV}$ and $\frac{d\mathcal{E}_{TOTAL}}{d\sigma}$ differ in signs. At the same time, the increasing functions of volume become decreasing functions of σ and vice versa. As a result, $\frac{d\mathcal{E}_{TOTAL}}{dV}$ and $\frac{d\mathcal{E}_{TOTAL}}{d\sigma}$ remain increasing functions. Fig. 5-1 and Fig. 5-2 confirm the above reasoning.

Thus, in each of the cases considered above, it is sufficient to know the dependences $\frac{d\mathcal{E}_{TOTAL}}{d\sigma}$ at some fixed value of σ_0, at which point it is supposed to determine the stability of the stationary state. After that, a graph of $\frac{d\mathcal{E}_{TOTAL}}{d\sigma}$ should be obtained, from which the nature of the resulting function can be determined. If the function is increasing, then the steady-state is stable. By repeating this procedure for different values of σ_0, the region of stability can be determined.

Eq. (5.36) presents the equality of two types of pressure at a steady state. The first term is positive and corresponds to the pressure of the film that tends to decrease its surface. The second term is the air pressure that is negative and tends to increase the air volume. For stability of the steady-state, it is required from a physical point of view that the first term should be greater than the second one with increases in the volume or with increasing β. In this case, the positive pressure arises that tends to decrease the volume. This pressure surpasses the pressure that tends to increase the volume. Eq. (5.36) contains an analytical expression that confirms this situation.

The energy of the compressed air is always a decreasing function of the volume. To provide a minimum of the total energy at a steady-state, another type of energy that compensates the air pressure (energy of a soap film, or stretched rubber shell, or energy of circulating light) should be the increasing function of the volume. These conclusions are also valid in the general case, although at high pressures the dependence of the pressure on compression requires adjustment.

We have presented these considerations in order to compare the stability of a soap bubble with the stability of a ball light. From the presented consideration, we can conclude that the existing methods for the analysis of the stability of the system have been developed very well.

In our particular case of circulating light, we need to analyze the dependence of the derivative of the total energy over volume. We have shown that the energy of the compressed air is a decreasing function of its volume. The derivative of the energy of compressed air by volume is always negative and has a simple physical meaning. This is air pressure with a minus sign. Therefore, the derivative of another energy over the volume must be positive. There is no soap shell in the circulating light. The shell is absent, but there is intense light circulating in the shell of compressed air. The derivative of the energy of light over the volume should be positive. It's hard to believe, but the existence of ball lightning in nature inspires optimism.

Steady states and stability of ball lightning

As has been shown for a soap bubble and a child's balloon, a stable steady state of an isolated system takes place if the derivative of total energy over the volume is an increasing function of volume. Condition Eq. (5.2) for the steady state of a soap bubble in the considered case is transformed as follows by

$$\frac{d\mathscr{E}_{TOTAL}}{d\sigma} = \frac{d\mathscr{E}_A}{d\sigma} + \frac{d\mathscr{E}_L}{d\sigma} = 0 \tag{5.38}$$

The only difference between Eq. (5.2) and Eq. (5.38) is that instead of the energy of the soap film $\mathscr{E}_S$ in Eq. (5.2), the energy of the circulating light $\mathscr{E}_L$ is used. True, these energies differ by many orders of magnitude, but this is a quantitative difference. Furthermore, while the compressed air is located within the volume confined by the soap film in a soap bubble, the compressed air is located within the thickness of the shell where intense light is circulating. A cross-section of the shell of the circulating light is shown in Fig. 5-3.

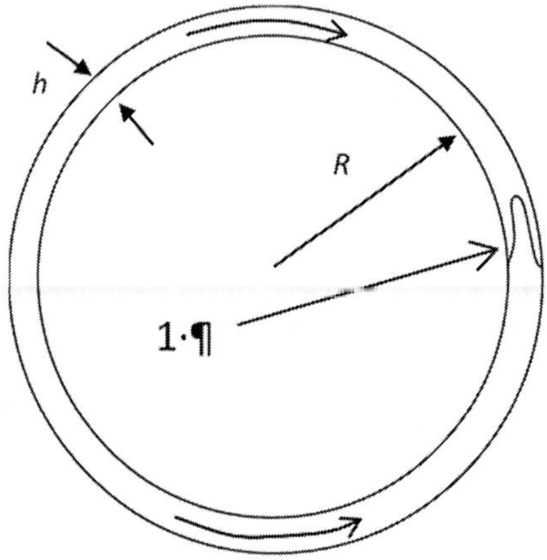

Fig. 5-3. Cross-section of a ball light. 1- Graph of dependence of the air pressure on the thickness.

Electromagnetic contra-pressure produced by circulating light

It seems obvious that a thin spherical film of compressed air where intense light is circulating in all possible directions cannot be stable. For a long time, we have been trying to prove that there are parameters of compressed air and circulating light for which such a combination is stable.

Following the same methodology that was used for the soap bubble, we analyze the behaviour of a sum of derivatives by volume (or by radius) from the energy of compressed air and the energy of circulating light. As already shown, the derivative of the energy of compressed air over the volume is negative, and its value is equal to the air pressure within the soap film.

Only about ten strings were required to analyze the dependence of the energy of the stretched soap film on the volume in the soap bubble. Since the dependence of the energy of the circulating light on the volume has not yet been studied by anyone, we consider this problem in a detailed way.

Let us take a thin spherical layer of compressed air of thickness h and radius R where intense white light is circulating in all possible directions. The cross-section of the sphere is shown in Fig.5-3. We have shown that the energy of light in an isolated conservative system in the form of ball light is in inverse proportion to the product of its radius and the reflective index of the compressed air (Torchigin 2003, 1-14). To verify this, just consider the simplest case where a light wave of a whispering gallery type circulates along the equator of a spherical shell. At resonance, the equator contains an integer number N of wavelengths. With a slow increase in the reflective index, the number N of wavelengths along the equator should be preserved. This property was justified early and follows from the fact that an abrupt change of the integer parameter N is impossible at a gradual continuous change of the product of the radius and reflective index.

The phase shift bypassing the equator is equal to $2\pi N$ where N is an integer. On the other hand, the phase shift is given by $2\pi R\omega n/c$, where ω is the eigenfrequency. At a slow change of parameters R or/and n, the integer N is not changed because N is an integer always and cannot change gradually. In this case, the product $R\omega n$ is not changed, and the following relation is valid

$$\omega n R = \omega_0 n_0 R_0 \tag{5.39}$$

where ω_0, n_0, and R_0 are the eigenfrequency, reflective index, and radius of the spherical shell, respectively, at a particular steady-state.

As is known from the theory of lossless closed oscillating systems, a change of the eigenfrequency at a slow change of the parameters of the system is in a direct proportion with a change of the energy stored in the system. The simplest explanation is that the energy of a photon is proportional to its frequency. Since the number of photons in a closed conservative system is unchanged, the energy of all photons is proportional to their frequency. This means that the total energy of an oscillatory system is proportional to its eigenfrequency. For the particular case of an oscillatory system in the form of an LC circuit, this property was analyzed in detail in Torchigin and Torchigin (2007, 167).

Thus, the energy of the circulating light $\mathcal{E}_L$ changes in a direct proportion with a change of the eigenfrequency ω. Then the following relation is valid

$$\frac{\mathcal{E}_L}{\omega} = \frac{\mathcal{E}_{L0}}{\omega_0} \tag{5.40}$$

Taking into account Eq. (5.39) and Eq. (5.40), a dependence of the energy light on parameters n and R can be presented as follows

$$\mathcal{E}_L = \frac{\mathcal{E}_{L0} n_0 R_0}{nR} \tag{5.41}$$

This simple equation is sufficient to analyze the derivative of the energy $\mathcal{E}_L$ over the radius R to get comprehensive information about the stability of the circulating light.

Function $\mathcal{E}_L$ is an increasing function of R if product nR is a decreasing function of R. In this case, the reflective index n should be a decreasing function of R. Indeed, with increasing R, the volume of the shell increases, the air density decreases accordingly, and therefore the reflecting index decreases also. Since the energy of the compressed air is a decreasing function of the radius, we need to show initially that the $\mathcal{E}_L$ can be an increasing function of the radius. As is seen from Eq. (5.41), this is possible only if a product nR is a decreasing function of the radius.

Unfortunately, the reflective index of the air is extremely small at normal conditions and is given by

$$n = 1 + \Delta n_N \tag{5.42}$$

where $\Delta n_N = 2.9\ 10^{-4}$.

With a change of the air density, only the small term Δn_N changes. As a result, a change of the air density, for example by 100 times, entails a change of the reflective index n by only several percent. It is difficult to expect that the product nR is a decreasing function of R. Nevertheless, light balls exist in nature, and we need to assume that the reflective index of the compressed air differs from unit to unit noticeably. It is possible only at extremely great air pressure in the shell. Thus, we can conclude without mathematical manipulations that the air pressure in the shell is extremely great. Then the derivative of the energy of the circulating light and, therefore, the energy of the circulating light should also be extremely great. The same conclusions have been obtained from explanations of the anomalous motion of natural ball lightning in the earth's atmosphere as well as from the extremely long life-time of circulating light (Torchigin 2019, 294-301).

We are now ready to show that the circulating light can be stable. As before, we will use the dimensionless parameter β given by Eq. (5.13).

It is impossible to predict in advance the behaviour of the thickness of the shell with a change of the radius R. There are various arguments concerning such behaviour. We choose the simplest and most reliable approach, in which the calculation is performed at several different values of this parameter. Modern numerical analysis tools allow you to perform calculations for several parameter values with the same success as for a single value. Based on the results obtained, it can be judged which value is closest to the truth.

Let us introduce the dimensionless parameter α that characterizes a dependence of the thickness of the shell with changing radius as follows

$$h = h_0 (R / R_0)^{\alpha} \tag{5.43}$$

where h_0 is the thickness of the shell at $R=R_0$. As has been seen, if $\alpha=1$, the thickness is proportional to the radius of the shell, if $\alpha=2$, the thickness is proportional to the surface of the shell, and if $\alpha=3$, the thickness is proportional to the volume that is confined by the shell.

In accordance with Eq. (5.41), the derivative of the energy of light over a volume at the state where $\mathcal{E}=\mathcal{E}_0$, $R=R_0$, and $n=n_0$ is given by

$$\frac{d\mathcal{E}_L}{dV} = -\frac{\mathcal{E}_{L0} n_0 R_0}{(nR)^2}\left(\frac{dn}{dV}R + n\frac{dR}{dV}\right) = -\mathcal{E}_{L0}\left(\frac{dn}{dV}\frac{1}{n} + \frac{1}{R}\frac{dR}{dV}\right) \tag{5.44}$$

The first term in Eq. (5.44) can be transformed as follows

$$-\mathcal{E}_{L0}\left(\frac{dn}{dV}\frac{1}{n}\right) = -\mathcal{E}_{L0}\left(\frac{dn}{d\rho}\frac{d\rho}{dV}\frac{1}{n}\right) = \mathcal{E}_L 0\left(\frac{dn}{d\rho}\frac{\rho}{V}\frac{1}{n}\right) = W_{L0}\frac{dn}{d\rho}\frac{\rho}{n} \tag{5.45}$$

where W_{L0} is the energy density of the circulating light given by

$$W_{L0} = \frac{\mathcal{E}_{L0}}{V_S} \tag{5.46}$$

Taking into account Eq. (5.43), the volume of the shell V_S can be presented as follows

$$V_S = 4\pi R^2 h = 4\pi R^2 h_0 (R/R_0)^{\alpha} = 4\pi R^{2+\alpha} h_0 (1/R_0)^{\alpha} \tag{5.47}$$

Taking into account Eq. (5.47), the second term in Eq. (5.44) is given by

$$\frac{\mathcal{E}_{L0}}{R}\frac{dR}{dV} = -\frac{\mathcal{E}_{L0}}{4\pi(2+\alpha)RR^{1+\alpha}h_0(1/R_0)^{\alpha}} = -\frac{\mathcal{E}_{L0}}{V_S(2+\alpha)} = -\frac{W_{L0}}{2+\alpha} \tag{5.48}$$

Then, Eq. (5.44) can be transformed as follows

$$\frac{d\mathcal{E}_L}{dV}=W_{L0}(\frac{dn}{d\rho}\frac{\rho}{n}-\frac{1}{2+\alpha}) \tag{5.49}$$

As seen from Eq. (5.49), the energy of the circulating light can increase with an increase of the volume if the expression in brackets is positive.

Eq. (5.49) is an analogue of Eq. (5.31), Eq. (5.21), and Eq. (5.6) for soap bubbles and children's balloons.

The dimensionless parameter $\frac{dn}{d\rho}\frac{\rho}{n}$ can be presented in the following form $\frac{(\Delta n)/n}{(\Delta\rho)/\rho}$. This is a ratio of a relative change of the reflective index to a relative change of the density. This ratio is extremely small for normal conditions. For example, if the air density increases by two times and $(\Delta\rho)/\rho=1$, then $(\Delta n)/n$ increases by $\Delta n=2.9\cdot10^{-4}$ only. Nature has found a way to increase this term so that the term in brackets becomes positive. It's hard to imagine, but circulating light can exist only in this case.

Dependence of the reflective index n of the air density can be determined from the following Lorenz relation

$$\frac{n^2-1}{n^2+2}=K\rho \tag{5.50}$$

where K is constant for each given substance.

Taking the logarithm of both parts of Eq. (5.50) and differentiating these parts, we have

$$d(n^2)[\frac{1}{n^2-1}-\frac{1}{n^2+2}]=\frac{d\rho}{\rho} \tag{5.51}$$

Eq. (5.51) can be transformed as follows

$$\frac{6ndn}{(n^2-1)(n^2+2)}=\frac{d\rho}{\rho}$$

or as

$$\frac{dn}{d\rho}\frac{\rho}{n} = \frac{(n^2-1)(n^2+2)}{6n^2} \tag{5.52}$$

We can obtain the same result by the same transformations from the Clausis-Mossotti relation.

Eq. (5.52) can be considered as a differential equation. We can get a dependence $n(\rho)$ through its numerical solution at the following initial conditions: air density ρ_0=1.225 kg/m^3, n_0=1.00029 at normal conditions where the air pressure is P_0=10^5 Pa and the air temperature is T_0=300 K.

The dependence $n(\rho)$ obtained at the numerical solution of Eq. (5.52) is shown in Fig. 5.4. .

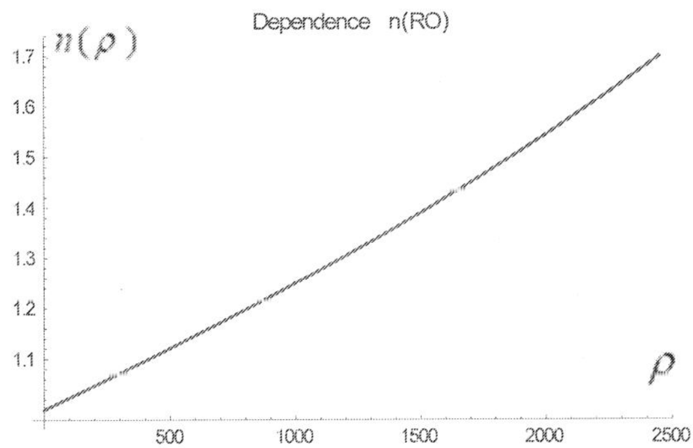

Fig. 5.4. Dependence of the air reflective index $n(\rho)$ where ρ is the air density ρ.

The air density ρ and its compression σ are related by the following relation

$$\rho = \sigma\rho_0 \tag{5.53}$$

The dependence of $n(\sigma)$ obtained from $n(\rho)$ and Eq. (5.53) is shown in Fig. 5.5.

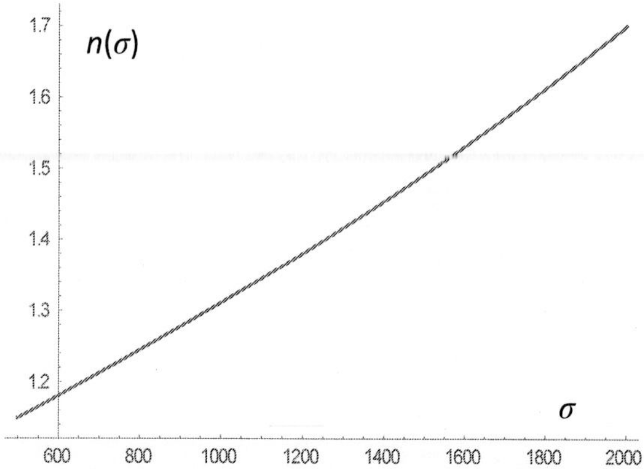

Fig. 5.5. Dependence of the air reflective index $n(\sigma)$ where σ is the air compression.

Having a known dependence $n(\sigma)$, we can determine from Eq. (5.52) the dependence $\frac{dn}{d\rho}\frac{\rho}{n}$ on compression, as is shown in Fig.5-6.

Taking into account Eq. (5.52), we can present Eq. (5.49) as follows

$$\frac{d\mathcal{E}_L}{dV} - W_L\left(\frac{(n^2-1)(n^2+2)}{6n^2} - \frac{1}{2+\alpha}\right) = W_L\gamma(n,\alpha) \tag{5.54}$$

where the dimensionless index γ is given by

$$\gamma(n,\alpha)=\frac{(n^2-1)(n^2+2)}{6n^2}\frac{1}{(2+\alpha)} \qquad (5.55)$$

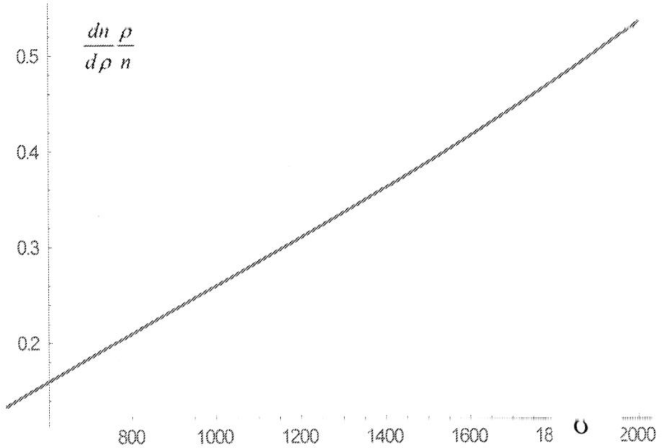

Fig. 5.6. Dependence of $\dfrac{dn}{d\rho}\dfrac{\rho}{n}$ on the air compression ρ.

This index deserves a separate name. We call it the γ-index. As noted in the analysis of the soap bubble, a prerequisite for the existence of a steady-state is the requirement that the function $\dfrac{d\mathcal{E}_L}{dV}$ must be positive and, therefore, the γ-index must be positive.

Dependencies of γ-index on the air compression σ for three different values of α are shown in Fig. 5.7. We can see from

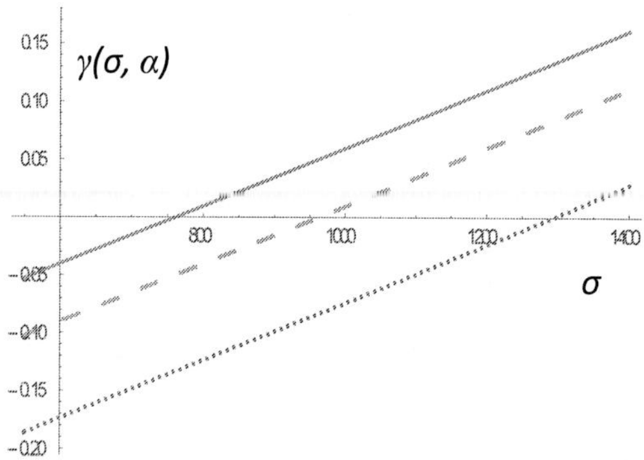

Fig. 5.7. Dependence of the derivative of the energy of light over the volume on the air compression for three values of parameter α.

Solid curve α=3.

Dashed curve α=2.

Dotted curve α=1.

The density of the compressed air at $\sigma>750$ is greater than $\rho>930$ kg/m^3 and is comparable with the density of water $\rho=1000$ kg/m^3.

We have refuted the main argument that the energy of circulating light always decreases with increasing radius. However, the question remains as to whether this steady state is stable. To answer this question, it is necessary to have information on the behaviour of the derivatives of the energy of the circulating light and compressed air in the vicinity of steady states. Now, we need to consider the behaviour of the air at these great compressions, where $\sigma>750$.

The air pressure at extremely great compression

We have considered the properties of the energy of circulating light based on the concepts and laws of optics. Another component of ball light is the compressed air at an extremely great pressure. There are no optics laws for this, but we need to analyze the properties of compressed air; otherwise, all our efforts to analyze the stability of circulating light do not make sense. The science of the behaviour of compressed gas originated more than two centuries ago. The Clapeyron equations for an ideal gas and the van der Waals equations for real gases are known. However, the compressed air in the shell is not a convention gas. When a gas is compressed more than 500 times ($\sigma>$ 500), the van der Waals equation ceases to be valid. Thus, the classical equations describing the state of gases are not suitable for our consideration.

The state of gas at pressures greater than 1 kbar (1000 atmospheres) is described by many different formulas with which researchers try to approximate their experimental results. You can find in Wikipedia the following models: the Redlich–Kwong model, Berthelot and modified Berthelot model, Dieterici model, Clausius model, Virial model, Peng–Robinson model, Wohl model, Beattie–Bridgeman model, and Benedict–Webb–Rubin model.

The presence of so many different models suggests that there is no universal model to reliably describe the state of gas at such high pressures. Only some specific experimental data for nitrogen in the pressure range from 1 to 22 kbar, collected from several sources in 0Tsiklis (1977), is reliable.

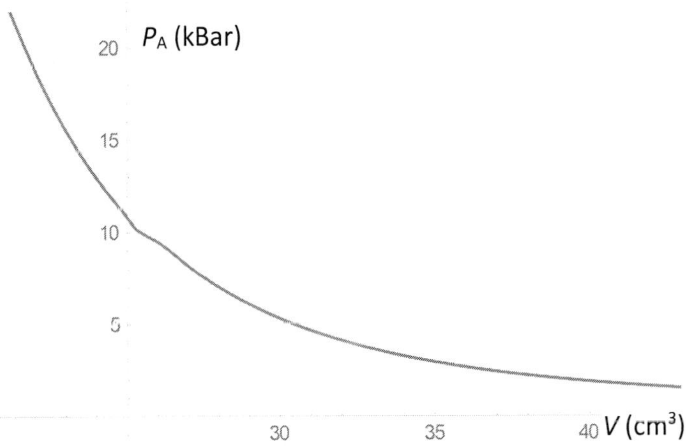

Fig. 5.8. Dependence of the gas pressure P_A (kBar) on volume V (cm^3) of 1 mole of nitrogen (22.4 liters at normal conditions) as derived from the experimental data.

These data are presented as the following list of 30 elements in which the pressure (in kbar) of one mole of nitrogen (22.4 litres at a normal pressure of 1 bar and a temperature of 300 K) is indicated when it is compressed to various volumes (in cm^3). The first value corresponds to the pressure in kbar; the second value corresponds to the volume in cm^3: {1.04, 47.68}, {1.533, 41.65}, {2.076, 38.02}, {2.563, 35.83}, {3.1, 34.05}, {4.07, 31.77}, {4.43, 31.13}, {5.15, 30}, {5.49, 29.54}, {6.27, 28.62}, {6.68, 28.21}, {7.09, 27.82}, {7.51, 27.46}, {7.96, 27.10}, {8.4, 26.76},

{9., 26.40`}, {9.5, 26}, {10, 25.47}, {11, 24.94}, {12, 24.44}, {13, 23.99}, {14, 23.58}, {15, 23.2}, {16, 22.86}, {17, 22.54`}, {18, 22.24`}, {19, 21.97`}, {20, 21.71`}, {21, 21.46`}, {22, 21.23`}

The graph derived from this list is shown in Fig. 5.8. Here, the dependence of the pressure of the gas pressure in kbar (10^8Pa) on the volume of one-mole gas (22.4 liters at normal conditions) is shown.

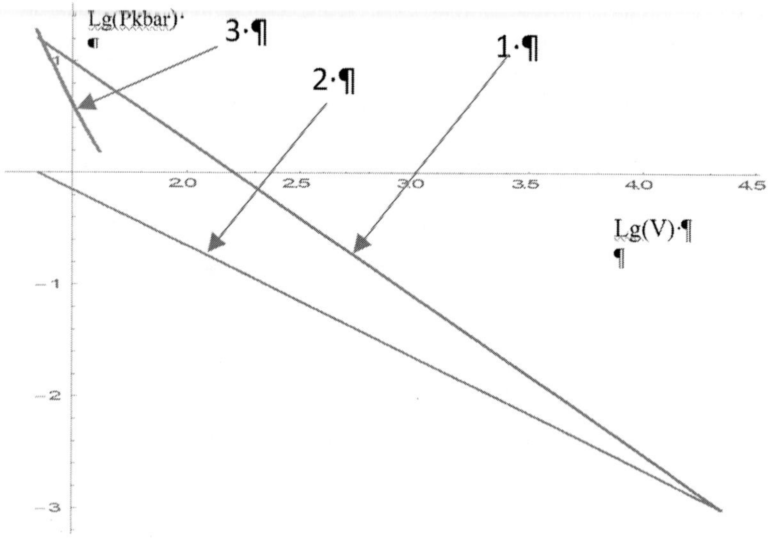

Fig. 5.9. Dependence of the air pressure on the volume of 1 mole on various assumptions.
The air is an ideal gas (curve 2).
The adiabatic process at the compression of the ideal gas (curve 1).
Experimental data (curve 3) at T=300 K.

The same graph in a logarithmic scale is presented by curve 3 in Fig. 5.9. Point 4.5 at the horizontal axis corresponds to $Lg_{10}(22400)$ where 22400 is the volume V_0=22400 cm^3 of one mole at normal conditions. $Lg_{10}(P)$ presents on the vertical axis the air pressure where P is expressed in kbar (1 bar corresponds to the normal atmospheric pressure about $P_0=10^5$ Pa). As can be seen, $Lg_{10}(P_0)$= -3. This means that $P_0=10^{-3}$ kbar or it is the normal pressure 1 bar.

Curve 2 shows the dependence of the pressure on the volume of an ideal gas in accordance with the Clapeyron law where $PV=RT$.

Curve 1 shows the dependence of the pressure on the volume for an ideal gas at an adiabatic process where $PV^{\gamma} = RT$.

Curve 3 shows the dependence of the pressure on the volume obtained from the experimental data shown in Fig. 5.8. As can be seen, the pressure increases with a decreasing volume much faster than for curves 1 and 2. This means that the compressibility of real air at extremely great compression is significantly smaller than that for an ideal gas.

As can be seen, the volume decreases by 1000 times from 22.4 litres to 22.4 cm^3. In this case, the air pressure increases from the normal pressure at $P_0=10^5$ Pa till 17.4 10^8 Pa, that is, by 17,400 times. This is 17.4 times more than follows from the Clapeyron equation, according to which the pressure is inversely proportional to the volume and should increase only by 1,000 times.

As is seen in the dependence Lg(P) on Lg(V) in accordance with curve 3 is close to a linear one. In this case, we can approximate this dependence by the following straight line

$$y - y_1 = \frac{y_2 - y_1}{x_2 - x_1}(x - x_1) \tag{5.56}$$

where x=Lg(V) and y=Lg(P). The coordinates $x_1 = 1.341$, $y_1 = 1.279$ and $x_2 = 1.609$, $y_2 = 0.187$ correspond to the upper and lower ends of curve 3, respectively. The volumes for the upper and lower ends of curve 3 are equal to V_1=21.97cm^3 and V_2=41.6 cm^3, respectively.

Passing back from Lg(V) and Lg(P) to V and P, and taking into account the compression $\sigma = V_0 / V$ where V_0=22.4 10^3 cm^3, we can present P in the considered region as follows

$$P = 10^{-10.548} \sigma^{3.93} \tag{5.57}$$

where P is expressed in kbar.

If pressure P is expressed in Pascal, Eq. (5.57) is transformed as follows

$$P = 10^{-2.548} \sigma^{3.93} \tag{5.58}$$

If pressure P is expressed in bars, Eq. (5.57) is transformed as follows

$$P = 10^{-7.548} \sigma^{3.93} \tag{5.59}$$

A comparison of the graph based on experimental data and the graph under Eq. (5.57) is presented in Fig. 5.10. As is seen, Eq. (5.57) satisfactorily describes the dependence obtained experimentally, removing the deviations associated with the scattering of experimental data.

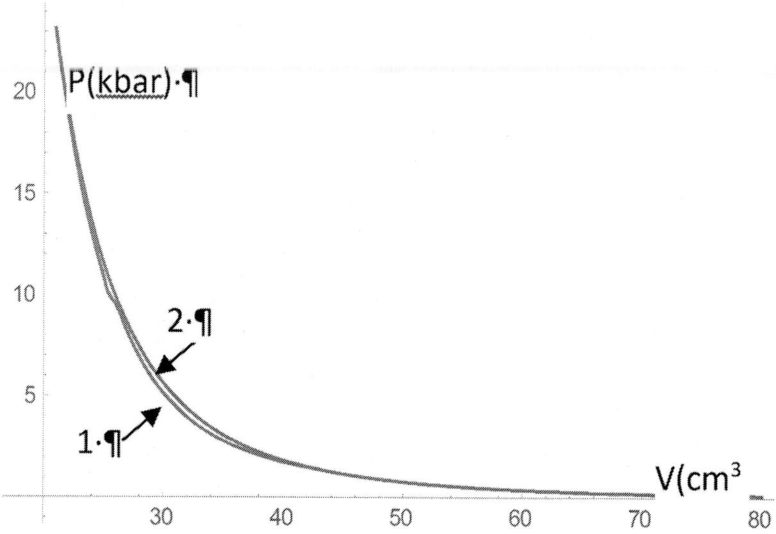

Fig. 5.10. Comparison of the dependencies P(V) under Eq. (5.57) (curve 1) and the experimental data (curve 2)

It can also be seen that the pressure is not proportional to the compression σ_N and increases faster than the compression increases.

The dependence $P(\sigma)$ of the air pressure over air compression derived from Eq. (5.58) is shown in Fig. 5.10.

We can assume that the segment of the straight line in Fig. 5.9 can be extrapolated in regions where $V>V_2$ and $V<V_1$. Then, Eq. (5.58) can be extrapolated in these regions.

We are also interested in the elasticity of highly compressed air. Usually, elasticity is considered for liquids and solids. Since steady-states of the circulating light are possible only at a sufficiently large compression $\sigma>\sigma_{TH}=750$, in which the density of compressed air is comparable to the density of water, it is useful to compare the behaviour of compressed air with the behaviour of liquids and solids known from the theory of elasticity under compression.

In the last case, the dependence of a change of a relative volume $\Delta V/V$ on the pressure P is given by

$$P = -K\frac{\Delta V}{V} \tag{5.60}$$

where K is the bulk module. For example, for water K=2.2 10^9 Pa.

Unlike liquids and solids, for which the bulk module is determined from the initial state, where the initial pressure is P=0, we need to determine the bulk module for the compressed gas in a state where the gas pressure is nonzero. Moreover, we would like to know the dependence of the bulk module on pressure.

Let us take the initial state P_0, ΔV_0. In this case, Eq. (5.60) looks as follows

$$P_0 = -K\frac{\Delta V_0}{V_0} \quad 5.61)$$

A change of ΔV_0 by δV is accompanied by a change of P_0 by δP. In this case, we have

$$\frac{\delta P}{\delta(\Delta V_0)} = \left.\frac{dP}{dV}\right|_{V-V_0} = -\frac{K}{V_0} \tag{5.62}$$

From Eq. (5.62), we have

$$K = -V_0 \left.\frac{dP}{dV}\right|_{V=V_0} \tag{5.63}$$

For example, we have for gases where the Clapeyron equation is valid

$$PV = Const \tag{5.64}$$

From Eq. (5.64), we have

$$\frac{dP}{dV} = -\frac{P}{V} \tag{5.65}$$

Then, from Eq. (5.63) and Eq. (5.65), we have

$$K = P \tag{5.66}$$

Thus, the bulk module of conventional gas at $\sigma<100$ is variable and is equal to the gas pressure. But we are interested in the bulk module of the compressed air at $\sigma > 750$.

Let us determine the dependence of the bulk module K on the air compression in this case. The volume of the compressed air is expressed as follows

$$V = V_N / \sigma \tag{5.67}$$

where $V_N = 22400\ cm^3$ is the volume of one mole of the air at normal conditions.

From Eq. (5.67), we have

$$\frac{dV}{d\sigma} = -\frac{V}{\sigma} \tag{5.68}$$

Then, Eq. (5.63) can be presented as follows

$$K = -V\frac{dP}{dV} = -V\frac{dP}{d\sigma}\frac{d\sigma}{dV} = V\frac{dP}{d\sigma}\frac{\sigma}{V} = \sigma\frac{dP}{d\sigma} \tag{5.69}$$

Taking into account and Eq. (5.58) and Eq. (5.69), the bulk module can be presented as follows

$$K = 3.93 \times 10^{-2.54}\sigma^{3.93} \tag{5.70}$$

Comparing Eq. (5.58) and Eq. (5.70), we can conclude that the bulk module of the strongly compressed air is greater by 3.93 times than the pressure of the strongly compressed air, unlike the air at small pressure where the bulk module of the air is equal to its pressure.

As seen in Fig. 5.11, the pressure of the compressed air is about 0.5 10^9 Pa at σ about 750. The bulk module of this air is about 2 10^9 Pa, and its density is about 900 kg/m^3. For comparison, the bulk module and density of water are equal to 2.2 10^9 Pa and 1000 kg/m^3, respectively.

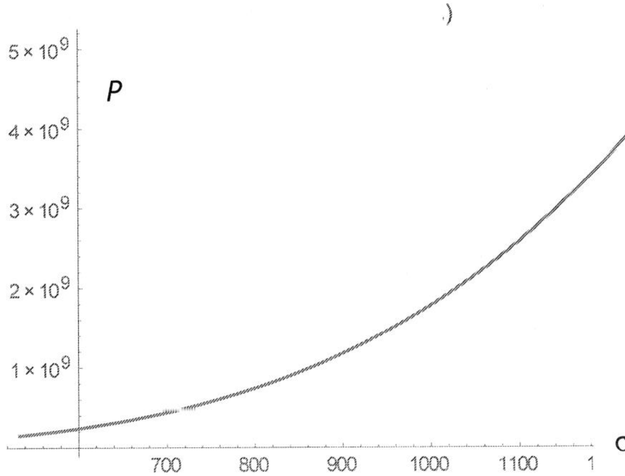

Fig. 5.11. Dependence of the air pressure P on the compression σ.

We are now ready to analyze the dependence of a derivative of the total energy of the circulating light and compressed air over the compression for different values of the parameter α.

Analysis of the steady state and stability of circulating light

As was shown, a steady-state takes place if a derivative of the total energy of an isolated system is equal to zero. There are two types of energy in a ball light. This is the energy of light, the derivative of which is given by Eq. (5.49), and the energy of the compressed air, the derivative of which is equal to the pressure of the compressed air and is given by Eq. (5.58).

In this case, the condition of stability of Eq. **(5.38)** can be transformed as follows

$$\frac{d\mathcal{E}_{TOTAL}}{dV}=\frac{d\mathcal{E}_L}{dV}-P_A=0 \tag{5.71}$$

Derivative $\frac{d\mathcal{E}_L}{dV}$ in accordance with Eq. (5.54) is the following

$$\frac{dE_L}{dV} = W_L \gamma(\sigma, \alpha) \tag{5.72}$$

The dependence of the air pressure on compression is given by Eq. (5.58). Thus, we have known dependencies of two terms in the right hand of Eq. (5.71) on the air compression σ. The same situation for a child bubble corresponds to Eq. (5.34).

As before, for a child balloon, let us analyze a steady-state for various values of the dimensionless parameter in the form of compression σ. Let a steady-state take place at arbitrary compression σ_0. A dependence of the air pressure P_A on compression σ, is known. Then a magnitude of the energy density of the circulating light at a steady state in accordance with Eq. (5.71), Eq. (5.72) is given by

$$W_L(\sigma_0) = P_A(\sigma_0) / \gamma(\sigma_0, \alpha) \tag{5.73}$$

In this case, an analog of Eq. (5.34) for the child balloon is the following equation

$$\frac{d\mathcal{E}_{TOTAL}}{dV} = W_L(\sigma_0)\gamma(\sigma, \alpha) - P_A(\sigma) \tag{5.74}$$

In analyzing the stability of a child balloon, we specifically demonstrated two approaches to the analysis of the stability of a soap bubble. The first approach, which is more understandable, is related to the behavior of a child balloon with an increase in its volume. It is shown that the volume derivative of the total energy in the steady state is equal to zero and is an increasing function of the volume. This means that the total energy in the steady state has a minimum and therefore the steady state is stable. In the second approach, another variable is used, the value of which is more abstract. This is a dimensionless value in the form of air compression σ, which, according to Eq.(5.67) decreases with increasing volume V. In this case, the derivative $\frac{d\mathcal{E}_{TOTAL}}{dV}$ is a decreasing function of σ. However, the sign of the derivative $\frac{d\mathcal{E}_{TOTAL}}{d\sigma}$ is different from the sign of the derivative $\frac{d\mathcal{E}_{TOTAL}}{dV}$.

Therefore, when the child balloon is stable, $\frac{d\mathcal{E}_{TOTAL}}{d\sigma}$ is an increasing function of σ. To confirm these considerations for a child balloon, whose stability is not in doubt, plots of dependences $\frac{d\mathcal{E}_{TOTAL}}{dV}$ and $\frac{dW_{TOTAL}}{d\sigma}$ in Fig. 5-2 and Fig. 5-1, respectively, were presented. As is seen, these dependencies are increasing functions of their arguments in both cases. Unlike volume which value can be various for different examples, compression σ is more pictorial variable. Besides the right part of Eq. (5.74) depends of the σ rather than on volume V. We can transform (5.74) as follows

$$\frac{dW_{TOTAL}}{d\sigma} = -\frac{W_L(\sigma_0)\gamma(\sigma,\alpha)}{\sigma} + 3.93\times10^{-2.54}\sigma^{2.93} \qquad (5.75)$$

where function $\gamma(\sigma,\alpha)$ is presented in Fig. 5.7.

Let a steady state takes place at a given sigma σ_0. In this case, the energy density at the steady state id given by

$$W_L(\sigma) = \frac{3.93\times10^{-2.54}\sigma^{2.93}\sigma}{\gamma(\sigma,\alpha)} \qquad (5.76)$$

When $\gamma(\sigma,\alpha)$ passes through zero, $\gamma(\sigma,\alpha)$ becomes positive, but close to zero. As a result, $W_L(\sigma_0)$ must be large so that the first term is equal in magnitude to the second. Thus, steady states are absent for compression values less than the value of σ, at which $\gamma(\sigma,\alpha)$ passes through zero.

The dependence of the energy density of circulating light $W_L(\sigma)$ in accoddance with (5.76) is shown in Fig.5.12.

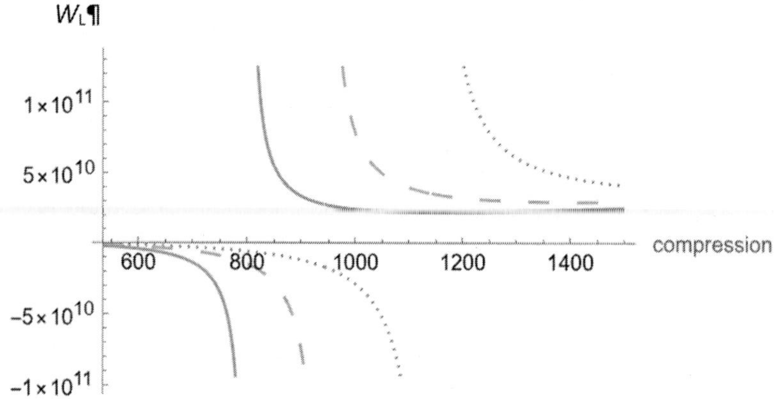

Fig.5.12. Dependences of parameter $W_L(\sigma_0)$ on compression σ for various parameters α, where solid curve corresponds α =3; dashed curve corresponds α =2; dotted curve corresponds α =1

Negative values correspond $\gamma(\sigma,\alpha)<0$. This indicates that the first term in Eq. (5.75) is positive and steady states cannot exist, since the second term is always positive.

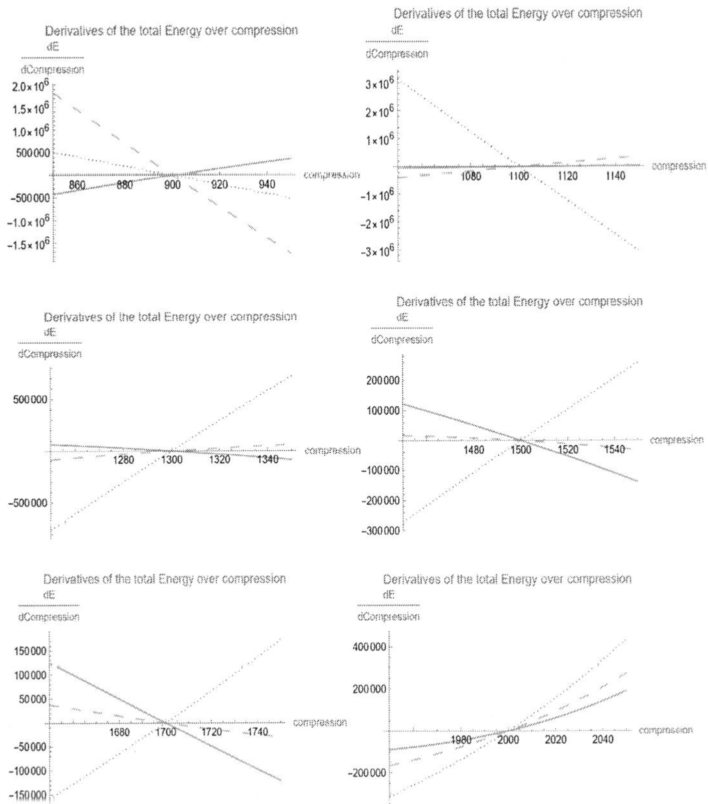

Fig. 5-13. Dependences of derivative of the total energy $\frac{dW_{TOTAL}}{d\sigma}$ over compression at various steady states determined by value of compression σ_0. Steady states take place at magnitudes of the compression σ_0 where the derivatives are equal to zero. Solid curves correspond α =3; dashed curves correspond α =2; dotted curves correspond α =1

As noted earlier at analysis of steady states of a child balloon, steady states are stable if the derivative of the total energy with respect to compression $\frac{dW_{TOTAL}}{d\sigma}$ is an increasing function of compression. It follows from the graphs in Fig. 5-13 that stable steady states are absent at σ <800. The steady states are stable for all values of the parameter a at σ >2000.

The relation between energies of light and compressed air

From Eq. (5.73), we can conclude that the ratio of the energy density of light to the air pressure is equal to $1/\gamma(\sigma,\alpha)$. Analyzing curves 2 in Fig. 5.7, we can see that this ratio is in a range from 5 to 10. For us, the ratio of the energy density of light to the energy density of compressed gas is more interesting. Since the volumes of gas and circulating light are the same, we could get an idea of the ratio of the energy of circulating light to the energy of compressed gas.

As is known, the energy density of the compressed air is equal to its pressure. However, this is true provided that the Clapeyron equation is valid. Let us assume that the dependence of the air pressure on the air compression σ at great pressures is given by Eq. (5.58) and calculate the energy $\mathcal{E}$ that is required to compress the air from σ=100 to σ=1000 or from 10^7 to 10^8 Pa.

The air energy $\mathcal{E}$ can be presented as follows

$$\mathcal{E}=\int_{V1}^{V2}-P(V)dV \tag{5.77}$$

Since $\sigma=V_0/V$ we have

$$\frac{dV}{d\sigma}=-\frac{V_0}{\sigma^2} \tag{5.78}$$

Eq. (5.58) can be presented as follows

$$P=a\sigma^b \tag{5.79}$$

where $a=10^{-2.548}$, b=3.93

Taking into account Eq. (5.78) and Eq. (5.79) we can present Eq. (5.77) as follows

$$\mathcal{E}=\int_{\sigma1}^{\sigma2}P(\sigma)\frac{dV}{d\sigma}d\sigma=\int_{\sigma1}^{\sigma2}P(\sigma)\frac{V_0}{\sigma^2}d\sigma=\int_{\sigma1}^{\sigma2}a\sigma^b\frac{V_0}{\sigma^2}d\sigma=\int_{\sigma1}^{\sigma2}aV_0\sigma^{b-2}d\sigma \tag{5.80}$$

Integrating Eq. (5.80) and taking into account that $\sigma_2>>\sigma_1$, we obtain

$$\mathcal{E}=\frac{aV_0(\sigma_2^{b-1}-\sigma_1^{b-1})}{b-1}\cong\frac{aV_0\sigma_2^{b-1}}{b-1} \tag{5.81}$$

The volume at compression σ is given by

$$V = \frac{V_0}{\sigma} \tag{5.82}$$

Then the air energy density $\mathcal{E}_D$ at compression σ_2 is given by

$$\mathcal{E}_D = \frac{aV_0\sigma_2^{b-1}}{b-1}\frac{\sigma}{V_0} = \frac{a\sigma_2^b}{b-1} = \frac{P(\sigma_2)}{b-1} \tag{5.83}$$

We can see from Eq. (5.83) that the energy density at extremely great pressure P is smaller about 3 times than the pressure P. Then we can conclude that the energy of the circulating light is greater than the energy of the compressed air by a factor in the range 15-30.

We have shown that the circulating light is stable with respect to changes in air compression in the shell. Any deviation of the radius or thickness of the shell from stationary values causes a change in the volume of the shell and therefore, a change in air compression. From the existence of a stable steady-state, it follows that there will be processes that will contribute to the restoration of compression and therefore, the elimination of these deviations. Thus, the circulating light is stable relatively changes in radius and thickness of the shell.

It is interesting if there is an upper limit for air pressure and energy density of circulating light, above which ball light ceases to be stable. We have reliable data on the dependence of air pressure on compression at σ=1100 only. The presented formula Eq. (5.58) based on these data shows excellent agreement with the experimental data shown in Fig. 5.10. It can be assumed that the formula is correct at extrapolation to higher compressions.

Another formula Eq. (5.52) should be taken into account also. The formula describes the dependence of the reflective index on the air density. Ultimately, this formula determines the dependence of the contra-pressure of circulating light on compression and the possibility of circulating light to confine the compressed air. We do not know the limits of the applicability of this formula. It is only known that the formula was used to calculate the reflective indexes of liquids where the compression is about 1000 (Ellingsen, Brevik 2011). The density of the compression air in this case is 1225 kg/m^3.

Another factor that affects the ultimate characteristics of circulating light is the dependence that determines a change in shell thickness with a change in the shell radius. In any case, based on reliable data on the behaviour of air pressure and the reflective index during compression in the range of 900 - 2000, we can see in Fig. 5-13 that there are stable steady-states of circulating light that are determined parameters of the pressure of compressed air and the energy density of circulating light, as well as the ratio between them.

We have presented evidence of the stability of the ball light using the same approach as was used for the analysis of the stability of a soap bubble. The same concepts, notation, and analysis methods were used in these two cases. In exactly the same way, the stability condition was determined by finding the minimum total energy. In both cases, only 2 types of energy participate in the balance. Moreover, one type of energy in the form of compressed air is common in both cases. Another type of energy in a soap bubble is the mechanical energy stored in the soap shell during its deformation. Another type of energy in the circulating light is the energy light presented by electrical and magnetic fields within the compressed air. Thus, from the point of view of physics and mathematics, the soap bubble and the circulating light differ little from each other. The difference is more quantitative than qualitative. No one doubts the existence of soap bubbles, although the task of calculating the minimum thickness of a soap film at which a soap bubble becomes unstable is no less complicated than calculating the minimal value of the energy density of the circulating light.

The problem in explaining the physical nature of ball light is that there is not currently a generally accepted theory that describes the mechanical properties of electromagnetic radiation. For more than 100 years, the debate on the magnitude of the momentum of light in an optical medium has been going on, although as early as 1904 Thomson would calculate the pressure of light on a flat boundary between free space and an optical medium (Thomson J.J. 1904 *Electricity and Matter*, Constable, London). The dispute is harmful because any work on the subject can be questioned since it is based on the grounds on which the dispute is being fought, and there is no generally accepted opinion.

Conclusion

To introduce the reader to the circle of those problems and concepts that are used in the analysis of any isolated conservative system, we considered initially examples where the stabilities of a soap bubble and child balloon are analyzed. The stability of the soap bubble is beyond doubt. We introduced approaches, physical concepts, and designations for the physical parameters that are necessary for the analysis and performed such an analysis. We explain such notions as a steady-state, stable steady-state, conditions of stability of the steady-state, total energy of an isolated system, different types of energies in the system, minimum of the total energy, derivative of the total energy over various parameters, pressure of the compressed air as a negative derivative of the energy of the compressed air. Then we present formulas that describe the conditions of stability and parameters of the system at a stable steady-state for a soap bubble.

The same concepts, methods and parameters were used at the analysis of the stability of the circulating light. Differences are more quantitative than qualitative. The only difference is the type of energy that is responsible for the confinement of compressed air. Unlike the mechanical energy of a stretched soap film, the energy of the circulating light is used in the form of the energy of alternating electric and magnetic fields of the light circulating in the compressed air.

Conditions of stability of the circulating light were analyzed. It is shown that stability is possible at an extremely great compression of the air that should be greater than 800. The reflective index of the compressed air should be greater than 1.3. The air pressure, in this case, should be greater than 1 kbar. The same conclusion is derived from the analysis of the lifetime of the circulating light. Dependencies between energies of the circulating light and the compressed air are determined at various compression of the air. It is shown that the energy of the circulating light can surpass the energy of the compressed air by a factor in the range 15-30.

Based on experimental data on the dependence of gas pressure on the compression ratio in the range 500–1000, a formula is obtained that correctly describes the behaviour of the gas in this range. This formula was used to calculate the parameters of circulating light and compressed air at a stable steady-state, as well as to calculate the ratio between the energies of circulating light and compressed air.

Analyzing the physical laws and the mathematical apparatus used in the considered theory of ball lightning or its mathematical model in the form of circulating light, we can conclude that such a theory could have been presented at a time when the concepts of force and energy were introduced by Newton and Joule, and also the foundations of mathematical analysis developed by Leibniz and the same Newton was known. No up-to-date information is required on the theory of relativity, quantum mechanics, nuclear physics, electrodynamics or plasma theory. It is enough to have knowledge of physics and mathematics within the scope of a comprehensive school.

It is shown that the circulating light is stable in a wide range of its energy density. As a result, the circulating light remains stable with gradually decreasing in time its energy. Since the pressure of the compressed air in the shell is gradually decreasing, there is a gradual decrease in the air density, the air reflective index, and an increase of the radiation losses. When radiation losses become greater than the losses due to the light scattering, the rate of the radiation losses increases sharply, circulating light becomes unstable and disappears abruptly and traceless.

Other parameters of the circulating light so as to total energy, the relation between energies of the light and compressed air, the mass of the circulating light, its deformity, possible velocities in the terrestrial atmosphere, reflective index of the compressed air are presented.

It is shown that a distinctive feature of ball light is that it can exist at extremely high air pressure and energy density of the light circulating in the shell. In this case, the air must be compressed approximately 1000 times, and the energy of the circulating light is approximately 15-30 times greater than the energy of compressed air. These results were obtained on the assumption that the dependence of the air pressure on compression at a compression of more than 1000 can be obtained by extrapolating the dependence in the range of 500 - 1000, and the dependence of the reflective index on compression obtained from the Lorentz-Lorenz law is valid for compression of more than 1000.

CHAPTER SIX

PHYSICAL PARAMETERS OF BALL LIGHTNING

In every branch of knowledge, the progress is proportional to the number of facts on which to build, and therefore to the facility of obtaining data.
J. Maxwell

Having ascertained the existence of steady states of circulating light, and their stability, we can begin to deduce the physical parameters of circulating light. From observations, it is known that the observed physical parameters of natural ball lightning can vary significantly, as do the parameters of soap bubbles. This applies to size, lifetime, stored energy, travel speed, and features of their disappearances. From the above discussion, it should be concluded that, in fact, there are practically no restrictions on the greatest magnitude of these parameters. We can only consider the phenomena that determine these quantities

At the same time, the very fact of the existence of ball lights imposes quite justified restrictions on the minimal values of these parameters. First of all, this concerns two main parameters. This is the magnitude of compression of compressed air and the energy density of the circulating light.

If there was no confirmation of the existence of circulating light in nature in the form of natural ball lightning, the magnitude of the air pressure and the energy density of light in the ball light could not even be mentioned in science fiction.

The pressure of the compressed air in the shell of the ball light

We have shown that ball light can exist only if the degree of air compression σ in the spherical shell exceeds a certain limit σ_{TH}, the value of which is extremely great and $\sigma_{TH}>800$ at least. In this case, the reflective index of the compressed air is above $n=1.3$, the density above 1225 kg/m^3, and the bulk module $K=2\ 10^9$ Pa. These parameters are comparable with those of water. The maximal magnitude of the air pressure in ball lights is difficult to estimate because we have information about the behaviour of the compressed air and the air reflective index at pressures not exceeding 2 GPa.

A simple extrapolation of the behaviour of graphs $n(\sigma)$ and $P(\sigma)$ to the region σ>1200 shows that there can be steady states and that they are stable.

It is known that at relatively low pressures when the Clapeyron law is valid, the energy density of the compressed gas is equal to its pressure. This is true for pressures of about 10^7 Pa or an air compression of σ=100. With further compression of the gas, the pressure grows faster than follows for the Clapeyron law. As a result, the energy density of the compressed gas is smaller than its pressure. It is shown that the energy density is about 3 times smaller than the air energy density in the region $10^2<\sigma<10^4$.

The energy density of the compressed air in the shell of the ball light

The air pressure in the shell of the ball light determines all other parameters, such as the energy density of the circulating light, the reflective index of the compressed air, and the centrifugal pressure.

For example, if the air pressure in the shell is Ps=10^9 Pa, then the energy density of the air is about 3 10^8 J/m^3 or 300 J/cm^3.

The energy density of the circulating light of the ball light

As has been shown, the energy density of light is greater than the energy density of air and can exceed it by two orders of magnitude. In this case, we can assume that the density of the total energy is about 3000 J/cm^3.

For comparison, a high-voltage 5 kV capacitor of capacitance C=100 mkF has a volume of 2,000 cm^3 and the energy stored within it is about 1,000 J. In this case, the energy density in the capacitor is 0.5 J/cm^3. This is 6,000 times smaller than the energy density in the ball light.

As another example, the energy density of a 1-kilowatt laser light beam of cross-section 1 mm^2 is 3J/cm^3, that is 1,000 times smaller than the density of the total energy in the ball light shell.

Energy stored in ball light

There are various estimates about the energy of natural ball lightning, starting from 1kJ if the lightning glows like a 100 Watt bulb for 10 s to 1 MJ if the ball lightning boils a barrel of water. Bearing in mind that the theory under consideration admits a wide range for the light energy density and the volume of the shell in which light circulates, such a spread in the energy of ball lightning is easy to explain.

For example, let the energy of the circulating light be $\mathcal{E}$=10 kJ = 10^4 J and the energy density be $W=10^{11}$ J/m^3. In this case, the volume of the shell is equal to $V=\mathcal{E}/W=10^{-7}$ m^3. On the assumption that the radius of the circulating light is equal to R=5cm = 0.05m, the thickness of the shell $h=V/(4\pi R^2)$ = 3.18 mkm.

This estimate is close to reality. Indeed, a spherical waveguide in the form of a film the thickness of which is h=3 μm, the reflective index of which is n=1.3 and the radius of curvature of which is R=5 cm, reliably confines the circulating light providing negligible radiation losses.

It should be noted that only two parameters cannot be obtained from observations of natural ball lightning. These are the compression index σ and shell thickness h. Changing these parameters provides a variety of observed ball lightnings. The parameter σ is limited from below and should be greater than 800. The parameter h is also bounded below by wavelengths of white light. The shell thickness can be less than one μm since the reflective index of the shell is relatively great and is equal to n=1.3. The extremely small thickness of the shell is evidenced by the fact that ball lightning can penetrate through thin slits while changing its shape.

The volume of ball light

It is necessary to distinguish between the volume of the ball light, which is limited by the shell of the ball light, and the volume of the shell itself. Since the air pressures inside and outside the shell are the same, the volume of the ball light should be understood as the volume of the shell. Assuming that the radius of the ball light is R=5 cm, and the shell thickness h=10 μm, we obtain that the shell volume is V – 3.14 10^{-7} m^3 or 0.314 cm^3.

Mass of ball light

The mass of the ball light consists of compressed air. Taking into account that the density of the compressed air is about 1000 kg/m^3, and that the typical volume of the shell and therefore that of the compressed air is $V=10^{-7}$ m^3= 0.1 cm^3, we obtain that the mass of the ball light is about m=0.1g.

It should be borne in mind that the air pressure inside the volume, which is confined by the shell, is equal to atmospheric pressure. Assuming that this volume is equal to 1000 cm^3, the mass of air is equal to 1 g. Thus, the weight of ordinary air in the ball bounded by a shell is about 10 times more than the weight of the shell itself.

The external force applied to ball light

The external force F that is applied to the ball light on the side of the external air atmosphere is equal to a product of the energy of the circulating light and the gradient of the reflective index of the external air (Torchigin and Torchigin 2017, 196). On the assumption that the energy of the ball light is equal to 10^4 J and grad(n)=10^{-6} m^{-1}, we have the force F=10^{-2}N. If m=0.1g, the acceleration of the ball light in accordance with the formula $a=F/m$ is equal 10^2 m/s^2.

To determine what the temperature change at a distance of 1 m should be so that grad (n) = 10^{-6}, the temperature dependence of the reflective index n is given by

$$n = 1 + \Delta n_N \frac{T_0}{T} \tag{5.84}$$

where Δn_N=2.7 10^{-4} and T_0=300 K is the air temperature at normal conditions. As can be seen, a change of the temperature by 1 degree causes a change in the reflective index by Δn_N/300=0.9 10^{-6}. Thus, a change of the temperature by 1 degree per 1 m causes a force that is about 10 times greater than the weight of the ball light.

It follows from the above estimates that the optically induced force F is comparable to the weight of the ball light, provided that the air temperature changes by one degree at a distance of 10 m. Thus, the ball light is an extremely sensitive device that responds to the slightest temperature change, provided that the air pressure and its composition are the same in the area under consideration.

Since the optically induced force is proportional to grad(n), where the reflective index n depends not only on temperature but also on the pressure and composition of the air, in some cases, these parameters have a decisive influence on the magnitude and direction of the force F. For example, the ball light arising in a linear discharge between clouds moves down even though the temperature of the air rises. This is because the air pressure increases to a greater extent, which leads to the air density increasing with decreasing altitude.

The reason that allows the ball light to catch up with the flying airliner becomes clear. The airliner disturbs the distribution of the air pressure around itself. Directly near the leading edge of an airliner wing, the air pressure increases by approximately 20%. This creates grad(n), which is several orders of magnitude greater than the grad(n) that existed in the same place before the arrival of the airliner. The value of this gradient gradually decreases with increasing distance from the airliner. But the grad(n) is always directed towards the airliner, where the grad(n) is maximum. Thus, the flying airliner creates the area around itself, where grad(n) is directed towards the airliner. This area moves with the airliner. Once in such an area, the ball light moves towards the airliner and is located in the place where the air pressure is maximum.

The lifetime of ball light

The lifetime of circulating light is greater by 4 orders of magnitude than that in the conventional air atmosphere. A decrease in the intensity of light propagating in the atmosphere is explained by the molecular scattering of light on fluctuations of air density (Landsberg 1976). The index of light scattering α in the air atmosphere is equal to α=2.7 10^{-7} cm^{-1} or 2.7 10^{-2} Km^{-1}. In this case, the intensity decreases by 15 times at a distance of 100 Km or at the time of about 3 ms. Analyzing a dependence of α on the pressure, we have shown that α decreases at extremely great pressure when fluctuations of the air density tend to zero. Index α decreases by 4 orders of magnitude at the energy density of light about 25 $10^{10}J/m^3$ (Torchigin 2003, 108-116). Thus, we concluded in 2003 from the explanation of the abnormally long lifetime of ball light that the densities of energies of the ball light and compressed air should be extremely great and far exceed the values that science and technology accepted.

Later, we reduced this estimate by an order of magnitude, since we took into account the following fact. Unlike light scattering in the Earth's atmosphere, in which scattered light is irretrievably lost, most of the scattered light in the shell of the ball light remains in the shell but continues to propagate in a different direction. This phenomenon helps to equalize the intensity of light and the directions of its propagation in each region of the shell and in all directions of propagation.

Electrical strength of the compressed air

When considering the electric strength of highly compressed air, which can withstand the total strength of the electrical field of the intense light circulating in the shell in all possible directions, it is necessary to keep in mind the following. The circulating light is a superposition of various waves with different frequencies, amplitudes, phases, and directions of polarization. All these circumstances lead to an increase in the ultimate density for intense white light if this concept can be applied. At present, there are no theoretical and experimental data on the limitations of the density of incoherent light in highly compressed air. Moreover, there is information that continuous optical breakdown from laser radiation can occur only at certain parameters of compressed air (Rizer 1980). The continuous optical breakdown is generally impossible at a pressure of more than 30 atmospheres. There is no upper limit to the intensity of light above which breakdown occurs. The air in the circulating light is compressed much more strongly, so there is no reason for the continuous optical breakdown.

Deformability of a spherical shape of a ball light

One of the main problems that cannot be solved within the framework of the concept of ball lightning as any kind of gaseous formation is the explanation of the existence of ball lightning as a whole, which restores its spherical shape after various deformations. Assuming ball lightning is a circulating light, this property of ball lightning is easy to explain.

Really, as is known, the shapeless closed homogeneous shell having a surface of *S* area, being pumped up by gas, gets the form of a body whose volume is maximal. It is a sphere for the given surface of the shell. A shell of a football ball can serve as an example.

Two phenomena are responsible for the spherical shape of the circulating light.

First is the centrifugal pressure created by light circulating in the shell in all possible directions. The density of light energy in each region of the shell is the same since light is scattered more strongly, the greater its intensity. This leads to the fact that the light intensities in all parts of the shell are the same. As a result, centrifugal pressures on all sections of the shell are the same, and the shell takes the form of a sphere. As has been shown, a typical centrifugal pressure that exerts the circulating light on the lateral surface of the shell is about $10^4 – 10^5$ Pa.

Second is the excess air pressure P_A in a thin spherical shell. This pressure is also uniformly distributed within the thickness of the shell and gives it a spherical shape. The pressure can be recalculated in the pressure on the lateral surface of the shell in the same manner as the centrifugal pressure is recalculated into the negative pressure within the shell. As has been shown, the centrifugal pressure is comparable with the contra-pressure, which in turn is comparable with the air pressure P_A. Thus, the recalculation of the air pressure P_A within the thickness of the shell into the pressure applied to the inner lateral wall of the shell gives a result comparable with the centrifugal pressure. Thus, the total pressure responsible for the spherical form of the shell is comparable with the centrifugal pressure, that is, a fraction of the atmospheric pressure. The air pressure P_A within the thickness of the shell is increased by a factor R/h, that is, about 10^4.

That is why ball light is easily transformed under the action of the external optically induced forces arising at its penetration through small holes. If ball lightning is located in a space with an inhomogeneous reflective index, but this inhomogeneity is characterized in all space by an identical gradient of the reflective index, ball lightning moves along this gradient without transformation. If the gradient of the reflective index in the considered area is different in different points of the ball, various sites of the ball lightning's surface are subjected to the action of various forces, and the ball lightning's movement is accompanied by its transformation. This property of ball light allows it to penetrate through cracks, where the thickness is much smaller than the ball light diameter.

Possible speed of ball light

It is beyond question that the greater the gradient of the air reflective index, the greater the ball lightning speed. An airliner flying at a speed of about 900 km/h or 250 m/s produces an excess pressure of about 0.38 bars near the front edge of the wing. As a result, there is a great gradient of the air reflective index in the region near the airliner. As has been shown, the external force applied to the circulating light is proportional to the air gradient. The magnitude of the gradient decreases gradually with an increasing distance between the airliner and the circulating light. As a result, the airliner creates a region around itself in which the gradient of the reflective index is directed toward the leading edge of the wing. This area moves with the airliner. Ball lightning, having got into such an area, begins to move in the direction of the gradient, that is, towards the airliner.

Reaching the area with maximum reflective index, ball lightning is located in this area. For any deviation from this region, ball lightning enters the region in which the gradient is directed to the initial region with the maximum reflective index. Having settled in this region, ball lightning moves along with the airliner and cannot be blown away by the oncoming airflow, the speed of which is several times higher than the wind speed of the strongest hurricane.

Furthermore, we cannot exclude that the circulating light motion is accompanied by sucking air from the surrounding space by the front hemisphere and removing air by the rear hemisphere. In this case, the circulating light motion can be imagined as screwing circulating light into the inhomogeneous atmosphere. This approach has been considered, and the same conclusion was derived, in that all the conditions for circulating light to catch up with a flying airliner are present.

CHAPTER SEVEN

APPEARANCE, EXISTENCE AND DISAPPEARANCE OF BALL LIGHT

Knowing the nature of ball lightning, having examined the laws and phenomena responsible for its existence, and taking into account its physical parameters, we can consider situations in which ball lightning can appear due to a discharge of ordinary linear lightning. Initial consideration of this problem was carried out in Torchigin (2004). Now, it is possible to consider this problem in more detail, since it has become possible to supplement the initial consideration with references to the laws, formulas, figures, and other materials given in this book.

The number of electrical discharges between clouds is approximately 2 orders of magnitude greater than the number of electrical discharges between clouds and the ground. But the occurrence of ball lightning during electrical discharges on the ground is much better studied.

The main problem is to determine the conditions under which light, usually propagating in a straight line, begins to propagate along closed paths inside a thin spherical layer of compressed air. The analogy with the formation of a soap bubble from a soap film may be helpful. As you know, a soap bubble occurs when the soap film manages to take the form of an almost closed surface with a small hole next to the tube.

Arising of ball lights at strikes of conventional linear lightning

The phenomena responsible for the arising of ball lights at strikes of conventional linear lightning were considered in Torchigin (2004, 494-500). A usual lightning strike is accompanied by the formation of a cylindrical sound shock wave due to the large current in the lightning core. In this case, the air temperature increases up to 25000-30000 C, and the current duration is equal to about 100 μs. The air pressure at the leading edge of the shock wave attains 0.3 GPa. In accordance with Fig. 5.11, this pressure corresponds to the air compression by about 600 times. The cylindrical wave expands with a velocity of 1.6 10^4 km/s, which is higher than the speed of sound in the atmosphere by a factor of about 50. Under these conditions, the air reflective index in the edge is greater than n>1.2. It is quite sufficient to form a thin-film cylindrical optical waveguide that can confine completely a light circulating around the axis of the cylinder of 10 cm diameter that is the typical ball lightning diameter. The light undergoes 1000 revolutions around the cylinder axis per 1 μs and the cylinder radius increases by 1.6 cm. There is intense radiation from the air heated up to 25000 – 30000 C.

Let us first consider the propagation of light in a thin cylindrical film of radius R shown in Fig. 7.1.

Fig. 7.1. The trajectory of the spiral beam inside a cylindrical glass rod.

This film has a face through which light can be introduced. The introduced light propagates in a spiral, where the angle between the direction of light propagation and the axis of the cylinder is constant. At the same time, an integer number of wavelengths N is laid on one coil of the spiral. In this case, the spiral wave can be represented as a superposition of two waves. One of them rotates around the axis of the cylinder, and the other propagates along the axis of the cylinder.

Now, we need to take into account some mathematical manipulations to present this spiral wave. The length of the rotating wave in the plane perpendicular to the cylinder axis is determined as follows

$$\lambda_T = 2\pi R / N \tag{7.1}$$

where N is any integer.

The propagation constant of the rotating wave is given by

$$k_T = 2\pi / \lambda_T = N / R \tag{7.2}$$

The vector of the propagation constant is rotating around the cylinder axis and is perpendicular to the axis. Total propagation constant directed along propagation of light is given by

$$k_0 = 2\pi n / \lambda_0 \tag{7.3}$$

where

$$\lambda_0 = \frac{2\pi c}{\omega} \tag{7.4}$$

Here ω is the angle frequency of the light wave.

The direction of the total propagation constant coincides with the direction of the wave beam. If the wave beam propagates at an angle α to the axis of the cylinder, the transverse component of the propagation constant is given by

$$k_T = k_0 \operatorname{Sin} \alpha \tag{7.5}$$

Taking into account Eq. (7.2) and Eq. (7.5), we have

$$\frac{N}{R} = k_0 \operatorname{Sin} \alpha \tag{7.6}$$

Let the radius R of the cylinder gradually decrease with increasing distance along the axis as shown in Fig. 7-2. In this case, the term N/R increases and, therefore, the angle α increases also. This means that the distance between the adjacent turns gradually decreases. For some R, Sinα becomes equal to 1. With a further decrease in the radius, Eq. (7.6) ceases to be fulfilled, and the wave cannot propagate further. The wave is reflected and begins to propagate in the opposite direction. The point from which the wave is reflected is called the return point.

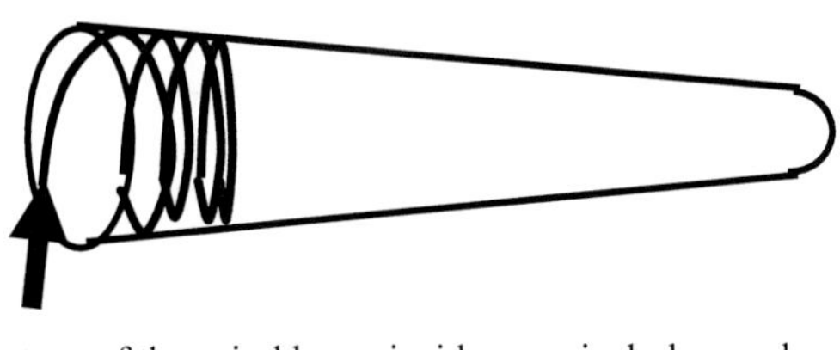

Fig. 7-2. Trajectory of the spiral beam inside a conical glass rod.

From Eq. (7.6), we have that the radius of the cylinder in the return point is given by

$$R_0 = \frac{N}{k_0} \tag{7.7}$$

The longitudinal component of the propagation constant is absent at the return point; therefore, taking into account Eq. (7.6) and Eq. (7.7), we have

$$\operatorname{Sin}\alpha = R_0 / R \tag{7.8}$$

Let us now consider a barrel-shaped segment of a cylindrical film, as shown in Fig. 7.3. In this case, different waves can successively reflect from two return points of the segment. It is easy to understand that the radiuses of the cylinder for the left and right return points are the same. In accordance with Eq. (7.8) radius R_0 depends on the angle α in the region where the cylinder radius R is maximal. The smaller this angle, the further the return points are removed from this section.

Consider now how the circulating light arises in the barrel-shaped cylinder. The circulating light aroused due to intense radiation from the air heated up to 25000 – 30000 C. The energy of the linear lightning is about 10^8 -10^9 J. The shell of the compressed air arises at the leading edge of the cylindrical shock wave where the air pressure is in the range 0.3 – 0.9 GPa. The air compression, in this case, is equal to σ=800, and in accordance with Fig. 5.5 the reflective index of the compressed air n=1.3. Of course, along with the linear lightning, the length of which can be several kilometers, several considered barrel-shaped shells can be formed. Let's consider one of them first.

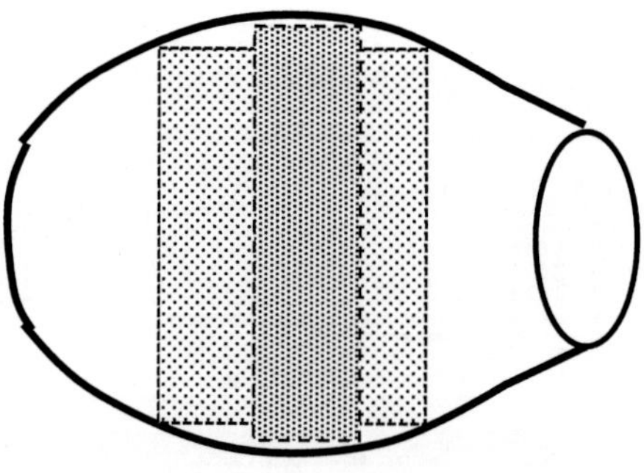

Fig. 7.3. Examples of regions of propagation of two types of circulating wave with various Sinα at maximal diameter. The shaded area corresponds to smaller Sinα.

Thus, we have the barrel-shaped cylindrical shell of the compressed air shown in Fig. 7.3, where very hot air radiates light in all possible directions. The bulk of the emerging light leaves the shell. However, a significant portion is confined with the shell and begins to circulate between the return points, as described above. All beams propagating along the tangents to the lateral surface of the shell will be captured by the shell if Eq. (7.8) is valid in the place where R is maximal.

It should be emphasized that in contrast to the beams that irrevocably leave the shell, the energy of the beams captured by the shell is gradually accumulated. As noted, within one microsecond, the radius of the shell increases by 1.6 cm. During this time, the light makes about 1000 revolutions, its energy density increases by 1000 times.

Besides, it is necessary to keep in mind the following additional method of energy concentration in the shell. The greater the Sinα, the smaller the distance between adjacent turns of the spiral, along which the light beams propagate, and, therefore, the greater the energy density of the light. As a result, we have great contra-pressure that tends to bring close to each other air molecules. This trend leads to the fact that there are conditions for reducing the diameter of the hole at the return point, where a distance between adjacent coils is equal to zero. A similar situation occurs when the hole in the soap bubble film is reduced at the moment when it is detached from the tube through which it was inflated.

When the holes disappear at both return points, a closed shell arises. The fact that ball lightning can differ from the ball shape and take any form is widely known. Some eyewitnesses testify that the ball lightning changed its form when it was penetrating through cracks and holes. Ultimately, the ball light acquires a spherical shape in an optical medium, where the air reflective index is unchanged at a distance equal to the diameter of the ball.

The initial length of the shell can be different and vary from a dozen centimeters to a dozen meters. The process of accumulation of the energy of light in the shell due to the emission of light by excited atoms into the shell can continue. We can imagine the following sequence of processes that can occur sequentially and/or simultaneously and lead to the accumulation of energy in a cylindrical layer with the subsequent formation of spherical ball light.

1. Accumulation of light energy in the cylindrical layer due to radiation of light excited by hot air. The atoms are excited due to the great air temperature produced by the shock wave in the time of discharge of the linear lightning.
2. Self-compression of the thickness of the cylindrical layer and concentration of light energy in the cylindrical layer due to a

decrease in its thickness owing action of the phenomenon of contra-pressure.
3. Collapse of two holes of the barrel-shaped cylindrical layer that leads to the formation of a spherical shell.

The first process occurs because the cylindrical layer with hotly excited atoms is optically transparent. The light energy density increases by a factor of 1000 times per microsecond because the radiation energy of heated air is added at each revolution to the energy of the light circulating around the axis of the cylindrical layer. This energy is accumulated in the cylindrical layer.

The second process is connected with features of propagation of light waves of whispering gallery type in the barrel-shaped region whose shape and thickness vary in time. The thickness decreases from several centimetres to several micrometers, i.e., by four orders of magnitude. The thickness of the layer becomes comparable with the light wavelengths. Since the volume where the light is propagating decreases, the density of light energy increases accordingly. The joint action of these processes can lead to a sufficient increase in the energy density of light in the barrel-shaped cylindrical layer.

Let us estimate the self-compression time to be sure that light is not strongly scattered in this time. Let us there are areas where the white light is produced by the electrical current of the linear lightning. Let the areas be located in an inhomogeneous medium in which the gradient of the reflective index is directed to the middle of a cylindrical layer and is equal to $g_n = 2\Delta n/h$, where $\Delta n \cong 10^{-1}$ is the maximal increase in a reflective index in a middle of the layer, $h \cong 3$ cm is the shell thickness. According to the eikonal equation $d^2h/dt^2 = 2g_n c^2$. In this case, peripheral sections reach the middle in time $t = c^{-1}(h/2g_n)^{1/2} \cong 0.4$ μs. This time is shorter significantly than the lifetime of light radiation in the air.

The third process that is connected with the collapse of two holes in a barrel-shaped cylindrical layer is not accompanied by a concentration of light energy but is rather important for ball light formation. It ought to be noted that the energy density of the light near the return points is greater than that near the maximal radius. Indeed, near the return points where $R=R_0$, Sinα is close to unity. This means that the distance between adjacent turns decreases as you approach the return point. In this case, the contra pressure near the minimal diameter is greater than that near the maximal one. As a result, the contra-pressure tends to decrease the radius of the hole at the return point. The same situation takes place in the soap bubble where the surface tension near the hole tends to decrease the hole's diameter.

We investigated the formation of ball light from one barrel-shaped segment. There can be several such segments in one linear lightning. These ball lightnings are located along with a bolt of linear lightning. Ball lightning can be formed in each segment. In this case, we get a one-dimensional structure of ball lightning. As the name implies, linear lightning is a one-dimensional structure. It seems that these segments are connected by a rope, just as the beads are connected by a thread. This structure is called bead lightning. From our consideration, it follows that the nature of bead lightning is the same as ball lightning.

Comparing the parameters of linear lightning (current, voltage, temperature, power, and pressure in the shock wave) with the parameters of laboratory installations, one can conclude that to obtain ball lightning comparable to natural ball lightning in terms of stored energy and lifetime, it is necessary to significantly increase the power and voltage of the gas discharge.

Abnormally great lifetime of ball light

As is known, the duration of the existence of natural ball lightnings lasts from several seconds to several minutes. This is significantly greater than the lifetime of conventional white light in the terrestrial atmosphere. Indeed, the sunlight decreases by 1.3% at a distance of 1 km in the terrestrial atmosphere. Bearing in mind that the speed of light is 300 000 km/s, we find that the light decreases 1.5 times per 100 μs. This is 4-5 orders of magnitude smaller than the lifetime of ball lightning. It is known that the phenomenon of molecular light scattering is responsible for decreasing the intensity of the light in the atmosphere.

We have shown in the very first work on the optical nature of ball lightning (Torchigin 2002, 108-116) that a decrease of the scattering light is possible at a significant decrease in fluctuations of the air density that are responsible for the scattering of light in the atmosphere. For this purpose, we have analyzed the following known expression for the intensity I of the scattering light (Landsberg 2003)

$$I = I_0 \frac{\pi V}{2\lambda^4 L^2}(\rho \frac{d\varepsilon}{d\rho})_T^2 \beta_T kT(1+Cos^2\theta) \qquad (7.9)$$

where I_0 is the intensity of light incident on the gas of volume V, λ is the light wavelength, L is the distance from the volume V to the point where the intensity of the scattered light is measured, and β_T is the isothermal compressibility of the gas given by

$$\beta_T = V^{-1}(dV / dP)_T \qquad (7.10)$$

where ρ, P, T, and ε are the density, pressure, temperature, and the dielectric constant of the gas, respectively, k is the Boltzmann constant and θ is the scattering angle of light. The single term that can be significantly decreased at extremely great air pressure measured by gigapascals is compressibility β_T.

The compressibility of the compressed gas can be presented as follows

$$\beta_T = 1/K \qquad (7.11)$$

where K is the bulk module.

Unlike liquids or solids, the bulk module is variable for gases and $K=P_0$ for the air at normal pressure. In accordance with Eq. (5.70), the bulk module at compression σ=1000 increases greater than 1 10^4 times. We can conclude from Eq. (7.11) that the compressibility of the compressed air at σ=1000 decreased by K_W/P_0=2 10^4 times. In this case, scattering of light in accordance with Eq. (7.9) decreases by 2 10^4 times and the lifetime increases accordingly.

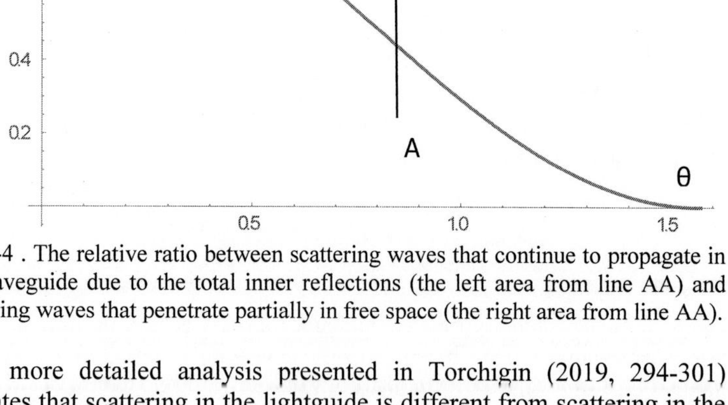

Fig. 7-4 . The relative ratio between scattering waves that continue to propagate in the waveguide due to the total inner reflections (the left area from line AA) and scattering waves that penetrate partially in free space (the right area from line AA).

A more detailed analysis presented in Torchigin (2019, 294-301) indicates that scattering in the lightguide is different from scattering in the air atmosphere because a majority of the scattering light continues to propagate in the lightguide due to the total inner reflection from sidewalls of the lightguide. By the way, the propagation of the main unrelated light is a series of reflections from the walls.

The dependence of the intensity of the scattering light on the angle of scattering θ is presented in Fig. 7-4 A vertical line separates the angles of scattering where the scattering light continues to propagate in the lightguide (left part) and the angles of scattering where the scattering light can be radiated in surrounding space. We need to note that in accordance with Fresnel's laws, only a part of this light leaves the lightguide. Another part returns to the lightguide due to the reflection. For us, it does not make sense to consider in more detail, since it is necessary first to determine the parameters of the lightguide from the compressed gas. As can be seen in, Fig. 7-4, the direction of the maximum intensity of the scattered light coincides with the direction of the main light.

Fig. 7-5shows the same dependence, but the horizontal axis presents the angle of scattering θ in radians. The area confined by two intersecting straight lines corresponds to the scattering light that continues to propagate in the lightguide. It is supposed that the reflective index of the lightguide surpasses by 20 percent the reflective index of surrounding space.

We can conclude that the difference between the molecular light scatterings in a homogeneous air atmosphere and the lightguide formed by the compressed air can be significant. A gas pressure of about 21-25 GPa was mentioned at an early stage, in Torchigin (2003, 104-108).

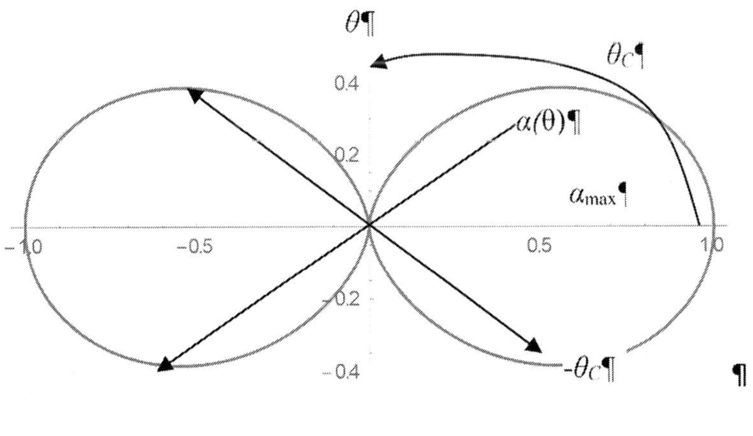

Fig. 7-5. Dependence of the scattering light on the angle θ between a direction of incident light and the direction of the scattering light. As is seen, the scattering is maximal along the incident direction.

From the presented consideration, this magnitude can be decreased to 2.1-2.5 GPa (Torchigin 2019, 294-301). This magnitude agrees with that obtained at the consideration of conditions of stability.

Features of the disappearance of the ball light

It is known from observations that ball lightning disappears unexpectedly, suddenly, and without a trace at any part of its trajectory. Sometimes it disappears silently, sometimes with a pop that resembles a pistol shot, that is, a sudden expansion of a small amount of gas.

In many ways, the disappearance of ball lightning is similar to the disappearance of a soap bubble. This concerns not only external signs but also the nature of the disappearance. In both cases, the system becomes unstable. In a soap bubble, the soap film degrades over time, and its thickness decreases in the upper hemisphere. The film bursts when it cannot confine the air pressure inside it.

A gradual decrease in the energy of light stored in circulating light is accompanied by a gradual decrease in the air pressure in its shell. This leads to a decrease in the air density and its reflective index. As a result, radiation losses gradually increase monotonously. As long as these losses are negligible as compared to other losses, the increase in radiation losses does not affect the total losses. However, there comes a moment when the radiation loss becomes equal to the loss due to molecular light scattering. After this moment, positive feedback begins to act. An increase in radiation losses leads to a decrease in the reflective index. A decrease in the reflective index leads to a further increase in radiation losses, etc. Since the speed of processes associated with light is 6 orders of magnitude greater than the speed of processes associated with compressed air, the light radiates from the shell much faster than compressed air expands. As a result, the circulating light disappears instantly.

The remaining compressed air, having deprived of the contra-pressure on the side of the circulating light, begins to expand, in the same way as gases expand after a bullet flew out of the barrel of a pistol.

After the air has been expanded, no traces of ball lightning remain. Therefore, ball lightning will disappear without a trace.

Before the onset of an unstable state, no signs indicating the approach of the disappearance of ball lightning are observed. The composition of light radiated by ball lightning remains constant. In some cases, just before the disappearance, there may be a short-lived outbreak associated with an increase in radiation losses before extinction.

In other cases, when the ball lightning shell contains a mixture of gases, an increase in the reflective index requires less pressure of these gases. In this case, the disappearance of ball lightning can occur silently (or low-frequency sound cannot be heard by the ear). In this case, the vapor of the gases contained in the shell can leave a trace after condensing in the atmosphere.

CONCLUSION

This book shows the impossible - something that no-one could have imagined before, something that seems contrary to common sense and to all known laws of physics and optics. Indeed, we are trying to theoretically substantiate that in the ordinary terrestrial atmosphere, ordinary white light can create a stable trap for itself in the form of a kind of bubble of light, in which white light continues to propagate at a tremendous speed close to the speed of light, but not in a straight line, as it usually does, but along closed paths. This surpasses all fantasy and borders on a mockery of existing ideas. Of course, there was not a single person ready to substantiate such a statement, if not for one essential circumstance. The behavior of such a hypothetical bubble of light, whose existence cannot be imagined based on simple, well-known laws of physics and optics, miraculously coincides exactly with the no less amazing behavior of natural ball lightning in the Earth's atmosphere. Such behavior has been observed with amazement for many centuries and recorded in the form of numerous testimonies of eyewitnesses. Natural ball lightnings can be seen as experimental samples created by nature itself to show people that their knowledge of nature is limited and incomplete.

Thus, this seemingly stupid task from the point of view of modern ideas is a task with a known positive result. Unfortunately, since 2003, when the optical theory of ball lightning was put forward, neither physicists nor mathematicians have been able to solve this problem.

After the author had been busy developing computers for 40 years, he himself had to take up the justification of his theory. It immediately became clear that the current state of knowledge regarding two simple key ideas related to the momentum of light and the pressure created by it in an optical medium was unsatisfactory, even though these issues had been known about for more than a century and a half, and hundreds of publications had been published on them.

Therefore, before proceeding to justify the stability of a bubble of light, it was necessary to determine the laws that were required for this purpose. Only one chapter is devoted to the justification of stability, with four chapters devoted to the justification of the laws used. These laws concern electromagnetic waves, but they are derived for waves of an arbitrary type, and electromagnetic waves are considered a special case. The laws are based on generally accepted laws of physics and optics and are obtained without any assumptions about the physical nature of the forces and pressures under consideration.

When studying modern ideas about the interaction of light with matter, it turned out that over the past 100 years, these ideas have moved far back and at present there is not even a generally accepted idea of the magnitude of the light pulse in the optical center. Moreover, there is no generally accepted idea of the magnitude of the forces acting on the optical medium from the light propagating in it. The situation turned out to be worse than at the beginning of the 20th century, when Thomson published a book on these issues in 1904 (it was Thomson who discovered the electron). The situation was even worse with another problem related to determining the pressure that light exerts on the medium in which it propagates. There are generally erroneous ideas, especially for gaseous media. The first four chapters deal with the above problems. Of course, the presented information is valid not only for light bubbles but is of a general nature. Therefore, it can be included in reference books and textbooks.

The main question, without which further analysis of the forces arising from the interaction of light with matter loses its meaning, has been resolved. Currently, there is no generally accepted opinion about the magnitude of the momentum of light. Disputes concerning mainly the magnitude of the momentum of an electromagnetic wave continue. Without entering into a fruitless debate, we have investigated the momentum of arbitrary waves. It has been shown that their momentum varies inversely with the velocity of their propagation. For the particular case of a light wave, it follows from this result that the momentum of a light wave is proportional to the ratio of the velocity of light in free space to the velocity of light in an optical medium, i.e., it is proportional to the reflection index of the medium. This result was obtained without any involvement of the specifics of light waves. However, this result was not enough, since another thought experiment by Balazs is known, also based on the general laws of physics and optics, according to which the momentum of light in an optical medium must be less than that in free space.

A modified Balazs experiment is considered, in which, instead of a light wave, an arbitrary wave propagating in the corresponding medium is considered. Similar results are obtained for such a wave. However, the interpretation of these results is different. It is shown that in this experiment, it is not the absolute value of the momentum that is calculated but the change in the momentums between two mediums.

It is shown that the absolute change in the momentum is inversely proportional to the wave propagation velocity. This result completely coincides with the results obtained earlier on the basis of other thought experiments. Thus, all known thought experiments lead to the same result, that the momentum of the wave is inversely proportional to the speed of its propagation. Therefore, we believe that the age-old dispute about the magnitude of the momentum of light in an optical medium should be considered completed because no contradiction exists between all the thought experiments and the momentum of light in an optical medium increases by n times as compared with that in free space.

By agreeing on the results of this experiment with the previously obtained results regarding the momentum of an arbitrary wave, it is shown that any wave produces in the medium in which it propagates new forces, the type of which depends on the type of the energy that the wave carries. These new types of forces transfer the momentum from the wave to the medium where the wave is propagating. As a result, the wave tries to drag along the medium in which it propagates. For a sound wave propagating in air, the experimental confirmation of such a drag is presented. Various parameters characterizing an arbitrary wave are concretized for elastic and electromagnetic waves. The physical natures and reasons for the appearance of the forces produced by these waves are explained.

Formulas are given for calculating these new types of forces for waves of various physical natures, such as an elastic wave, a sound wave, a wave in a stretched string, a torsion wave in a thin-walled pipe, and a sea wave in shallow water. It is shown that these new types of forces are responsible for the movement of energy bundles of a traveling wave in the medium in which the wave is propagating.

As for the special case of an electromagnetic wave, these new types of forces by their nature coincide with the known Abraham forces, the existence of which in nature continues to be debated to this day. The new types of forces are compared with the forces considered in the approach based on the Lorentz forces. Although the nature of these forces is the same, their magnitude is different. Since new types of forces are obtained on the basis of generally accepted laws of physics without any preliminary assumptions, they, in our opinion, correspond to the truth. In addition, the formulas for these new forces turn out to be much simpler. Recommendations are given for correcting the formulas for calculating forces based on the approach using the Lorentz forces.

Even more particular is the case of a wave of an arbitrary type is an electromagnetic wave propagating in free space. It is shown that the same types of forces must exist in free space. This follows from Newton's laws, according to which a change in the momentum of any object in time must be accompanied by forces responsible for this change.

The phenomena responsible for the compression of matter in which an electromagnetic wave propagates are analyzed. It is shown that the currently used formulas require correction. This applies to several points. First, one should take into account the fact that the matter is not located in a vacuum, but is located at normal atmospheric pressure. Therefore, it is required to take into account the initial compression by this pressure. Secondly, when determining the balance between the energy of an electromagnetic wave and the energy of compression of a matter, one should take into account the total energy of an electromagnetic wave, and not just the energy of an electric field. Thirdly, the energy of a compressed gas should be calculated using different formulas from those used for the energy of liquids or solids. It is shown that the compression of a gas by an electromagnetic wave is possible only at an extremely high energy density of the electromagnetic wave, which is achievable only in circulating light.

Particular attention is paid to explaining the fact that a gas can be subjected to strong compression without the use of a vessel that limits its volume if intense light propagates through it. It is shown that this is possible in a thin spherical film in which intense light circulates in all possible directions, as was used by nature in ball lightning.

The results concerning the laws of interaction of light with matter were used for analysis of the stability of circulating light. Since the methods of analysis of the stability of closed systems are not directly related to electrodynamics, these methods were first used as an illustration in the analysis of the stability of objects whose stability is beyond doubt. A detailed analysis of the use of these methods in the analysis of the stability of a soap bubble and a children's balloon is given. In both cases, one of the types of energy is the energy of compressed air. For a soap bubble, another type of energy is the energy of a soap film. For a child balloon, another type of energy is the energy of a stretched rubber shell. For circulating light, one type of energy is also the energy of compressed air in a thin spherical shell. Another type of energy is the energy of circulating light.

Thus, the bubble of light differs from a soap bubble in only one type of energy, which must be an increasing function of the radius of the sphere. It has been shown that this is only possible if the reflective index of compressed air is high enough, which is possible only with sufficiently strong air compression, at which the known laws of the dependence of pressure on compression become unfair. On the basis of the known experimental data on air pressure at high compressions, the necessary dependences were obtained. The same problems concerned the dependence of the reflective index of compressed air on the air compression were solved. Based on the dependences obtained, an analysis of the stability of the circulating light was conducted. The necessary conditions to ensure the stability of the bubble of light are deduced.

It is shown that the bubble of light is stable in a wide range of its light energy density. As a result, the circulating light remains stable with gradually decreasing in time its energy. Since the pressure of the compressed air in the shell is gradually decreasing, there is a gradual decrease of the air density, the air reflective index, and an increase of the radiation losses. When radiation losses become greater than the losses due to the molecular light scattering, the rate of the radiation losses increases sharply, circulating light becomes unstable and disappears abruptly and traceless.

Other parameters of the bubble of light, such as the total energy, the relation between energies of the light and compressed air, the mass of the circulating light, its deformity, possible velocities in the terrestrial atmosphere, and the reflective index of the compressed air, are presented.

Thus, the analysis of phenomena in the world of circulating light, which until now has not been mentioned or considered by anyone, made it possible not only to explain the nature of ball lightning but also to gain additional knowledge about the laws of nature. Only the erroneous belief that there can be no circulating light can explain the inattention to this phenomenon, although the knowledge accepted a century and a half ago is sufficient for its understanding.

REFERENCES

Balazs Nandor. 1953. "The Energy-Momentum Tensor of the Electromagnetic Field inside Matter". *Phys. Rev.* 91, 411.

Barber Bradley, Hiller Robert, Lofstedt Ritva, Putterman Seth, Wennger Keith. 1997. "Defining the unknowns of sonoluminescence", *Phys. Rep.*, 281, 65, 67-141.

Barber Bradley. Putterman Seth. 1991. "Observation of synchronous picosecond sonoluminescence" *Nature* 352: 318-320.

Barnett Stephen, Loudon Rodney. 2010. "The enigma of optical momentum in a medium". *Phil. Trans. R. Soc.* 368 927.

Barry James. 1980. "*Ball lightning and Bead Lightning".* NY: Plenum Press.

Brevik Iver. 2009. Comment on "Observation of a Push Force on the End Face of a Nanometer Silica Filament Exerted by Outgoing Light". *Phys. Rev. Lett.* 103, 219301.

Eden Alec. 1992, "*The search for Christian Doppler*", Wien, Springer-Verlag.

Einstein Albert, Laub Jacob. 1908. *Ann. Physics*. 331, 541 or *Einstein's Collected Papers.* 1989. Vol. 2, Princeton University Press.

Ellingsen Simen, Brevik Iver. 2011. "Electrostrictive fluid pressure from a laser beam". *Phys. Fluids* 23, 096101 (2011);

Gordon James. 1973. "Radiation Forces and Momenta in Dielectric Media". *Phys. Rev. A.***8** No.1, 14.

Griffith David. 2012. "Resource Letter EM-1: Electromagnetic Momentum" *Am. J. Phys.* 80, 7.

Grigor'ev A.I., Grigor'eva L.D., Shiryaeva S.O. 1992. "Ball lightning penetration into closed rooms: 43 eyewitness accounts", *J. Sci. Explor*, 6(3): 261-279.

Grigoriev I. S. 1991 "*Handbook of Physical Quantities*", Moscow, Energoatomizdat.

Landau Lev, Lifshits E. M., and Pitaevski L. P. 1976. "*Electrodynamics of Continuous Media*". Oxford, Heinemann.

Landsberg, G. S. 1976. *Optics.* Moscow, Nauka.

Loudon Rodney, Barnett Stephen. 2006. "Theory of the radiation pressure on dielectric slabs, prisms and single surfaces". *Optics Express* 14 No. 24 11855.

Mansuripur Masud, Zakharian Armis, Wright Ewan. 2013. "Electromagnetic-force distribution inside matter". *Phys. Rev. A* 88 023826.

Mansuripur Masud. 2010. "Resolution of the Abraham-Minkowski controversy". *Opt. Comm.* 283, 1997-2005.

Mesyats G A, 2000. "Actones in a vacuum discharge: breakdown, spark". Moscow, Nauka.

Oraevskiy, A. N. 2002. "Whispering gallery waves". *Quantum electronics*, 32, № 5: 377–400.

Physical encyclopedia, 1992. Ed. Prokhorov A M vol. 3 p 32. Moscow, Bol'shaia Rossiyskaia encyclopedia.

Rizer Yu. 1980. "Foundations of modern physics of gas discharges". Moscow, Nauka.

Sagan Paul. 2004. "*Ball lightning: Paradox of physics*". NJ, Lincoln.

Singer Stanley. 1971. "*The Nature of ball lightning*". Plenum Press: NY.

Stakhanov Igor. 1979. "*The physical nature of ball lightning*". Atomizdat, Moscow.

Steinhoff Mark. 1998. "Ball lightning – An Unsolved Problem in Atmospheric Physics". Plenum PNY.

Thomson Joseph. 1904. *Electricity and Matter*. Constable, London.

Torchigin Vladimir. Torchigin Alexander. 2016. "Photon and physical phenomena responsible for its momentum". *Optik.* 127 6120-6128.

Torchigin Vladimir, Torchigin Alexander. 2007. "Self-organization of intense light within erosive gas discharges". *Physics Letters* A 361 167.

Torchigin Vladimir, Torchigin Alexander. 2010. "On phenomenon of light radiation from miniature balls immersed in water". *Physics Letters A*: 374-380.

Torchigin Vladimir, Torchigin Alexander. 2013. "Interrelation between ball lightning and optically induced forces". *Phys. Scr.* 88: 035402 (6pp).

Torchigin Vladimir, Torchigin Alexander. 2014. Comment on "Theoretical analysis of the force on the end face of a nanofilament exerted by an outgoing light pulse". *PRA* 89 no.5, 057801.

Torchigin Vladimir, Torchigin Alexander. 2016. "How the ball lightning enters the room through the windowpanes". *Optik*, 127: 5876.

Torchigin Vladimir, Torchigin Alexander. 2016. "How Ball Lightning penetrates in room through small holes and splits". *Optik*. 127 6155.

Torchigin Vladimir, Torchigin Alexander. 2017. "How ball lightning manages to catch up a flying aircraft and penetrate into its salon. *Optik*, 148; 196.

Torchigin Vladimir. 2003. "On the Nature of Ball Lightning". *Doklady Physics*, Vol. 48, No. 3:108-116.

Torchigin Vladimir. 2003. "Optical Resonators in the Atmosphere". *Laser Physics*, Vol. 13, No. 6, 1-14.

Torchigin Vladimir. 2004. "Mechanism of the Appearance of Ball Lightning from Usual Lightning". *Doklady Physics*, Vol. 49, No. 9, 494-500.

Torchigin Vladimir. 2018. "Explanation of abnormal behavior of ball lightning near the earth surface". *Optik*, 171 167

Torchigin Vladimir. 2018. "Explanation of abnormal motion of ball lightning near the earth's surface." *Optik*, 171: 149.

Torchigin Vladimir. 2018. "Explanation of anomalous bouncing luminous droplets of liquid silicon in a framework of the optical model of Ball Lightning". *Optik*, 171:188.

Torchigin Vladimir. 2019. "Ball lightning as a bubble of light: Existence and stability". *Optik*, 193: 162961.

Torchigin Vladimir. 2019. "Explanation of abnormal motion of glowing silicon balls in a framework of optical model of ball lightning" *Optik*, 176: 704.

Torchigin Vladimir. 2019. "How ball lightning finds out slots and holes to penetrate through them". *Optik*, 194: 163126.

Torchigin Vladimir. 2019. "How optical model of ball lightning explains its paradoxical movement upwind". *Optik*, 184: 533-537.

Torchigin Vladimir. 2019. "Why ball lightning disappears suddenly traceless". *Optik*, 187: 65-69.

Torchigin Vladimir. 2019 "Physics of a ball lightning in a form of a bubble of light". *Optik,* 194: 294-301

Torchigin Vladimir. 2020. "Physical phenomena responsible for stability and spherical form of ball lightning". *Optik*, 219:165098.

Torchigin Vladimir. 2021. "Sonoluminescence and circulating ligh". *Optik*, 239 (3): 166799.

Torchigin Vladimir. 2021. "Ball lightning and the Chernobyl accident". *Optik,* 241: 167224.

Torchigin Vladimir. 2021. "Behavior of the circulating light in a mixture of gases". *Optik*, 243: 167398.

Torchigin Vladimir. 2021. "Circulating light and a separation of components of a mixture of gases". *Optik*, 244: 167524.

Torchigin Vladimir. 2021. "Circulating light at continuous electrical current". *Optik*, 244: 167395.

Torchigin Vladimir. 2021. "Circulating light that burns out holes in metal sheets". *Optik*, 243: 167528.

Torchigin Vladimir. 2021. "Parameters of natural ball lightning derived from its optical model". *Optik*, 243: 167390.

Torchigin Vladimir. 2021. "Steady states and stability of the circulating light in the air atmosphere". *Optik*, 235: 166635.

Torchigin Vladimir. 2021. "Unknown features of a behavior of the circulating light". *Optik*, 241: 167233.

Turner David. 1998. "Ball lightning and other meteorological phenomena", *Physics reports*; 293, 1–60.

INDEX

AFTERWORD

We have tried to convince the reader that in addition to the familiar world with light surrounding us, there is another world, hitherto unknown, the world of circulating light, in which additional forces begin to manifest themselves. It is the presence of these forces that explains the anomalous behavior of luminous objects from this world.

The nature of ball lightning in the form of circulating light has been known for about 20 years. However, it cannot be said that the number of supporters of this theory is growing in the world. The famous Copernicus theory that the earth revolves around the sun required more than the century to gain universal acceptance. The new theory is usually not accepted by contemporaries. New generations, comparing existing theories, give preference to the most correct from their point of view.

The situation with circulating light is completely different. We have already shown that attempts to carry out a thermonuclear reaction based on the phenomenon of sonoluminescence are futile since the visible shift of the radiation spectrum to the short-wavelength region is associated with the compression of the circulating light, and not with radiation of the body heated to a very high temperature. In addition, we have shown that circulating light must be considered in high-risk situations such as flying airliners and nuclear power plants.

Besides, without waiting for universal acceptance, the world of circulating light allows a deeper understanding of known phenomena and a broader understanding of the laws of the world around us.

The most difficult task was to prove that the circulating light can be stable. In the very first article, we showed that circulating light can exist for tens of seconds if the air pressure in the spherical shell is extremely high and the air density is comparable to that of water. Further, this result was confirmed by studying the relationship between the energies of light and compressed gas.

We have just begun to study the phenomena in the world of circulating light. These phenomena require further study. For example, the world of nuclear physics was discovered at the end of the 19th century. Only half a century later, these phenomena have found practical application.